Edited by

G. Denardo and H. D. Doebner

Conference on Differential Geometric Methods in Theoretical Physics

International Centre for Theoretical Physics, Trieste
30 June - 3 July 1981

World Scientific

World Scientific Publishing Co Pte Ltd
P O Box 128
Farrer Road
Singapore 9128

ISBN 9971-950-58-8

Printed in Singapore by Richard Clay (S. E. Asia) Pte. Ltd.

FOREWORD

The Conference on Differential Geometric Methods in Theoretical Physics was held at the International Centre for Theoretical Physics in Trieste from 30 June to 3 July 1981. The directors were G. Denardo (University of Trieste) and H.D. Doebner (University of Clausthal, F.R.G.).

Besides the International Centre for Theoretical Physics, the International School for Advanced Studies in Trieste and the Office for Foreign Studies and Activities at the Technical University of Clausthal (Federal Republic of Germany) have given their contribution to the success of the conference. We are presenting here the lectures and seminars which have been given.

The Conference has been organized in the spirit of the tradition of the earlier conferences on the same theme held in Aix-en-Provence, Bonn, Warsaw, Clausthal and Salamanca. While focusing primarily upon recent advances in geometrical and topological aspects of field theories, non-Abelian gauge theories, supersymmetries, general relativity and quantization methods, the main aim was that of bringing together physicists and mathematicians representing a wide spectrum of interests.

The directors of the Conference wish to thank particularly Professor Abdus Salam, Director of the International Centre for Theoretical Physics, Trieste, and Professor Paolo Budinich, Director of the International School for Advanced Studies in Trieste for their essential scientific, moral and material support to the Conference.

G. Denardo H.D. Doebner

CONTENTS

I. GENERAL REVIEW LECTURES

ADVANCES IN SUPERGRAVITY

Yuval Ne'eman*,**
Tel Aviv University*** and University of Texas, Austin†,††,†††

1. The Physics of Supergravity

In recent years, General Relativity has re-entered Physics, with observations and theoretical developments influencing each other. Steady-State Theory was rejected following the discovery of the $3^{o}K$ Background Radiation; an active search for Black Holes is taking place in connection with the new understanding of Black Hole Physics; "Grand-Unification Gauge Theories" (GUTs) have been applied to Cosmogony, providing new (speculative) explanations for the matter-antimatter asymmetry and for the 10^8 ratio of photons to baryons in the Universe.

By its very nature, Supergravity should (if successful) provide in the long run the keys to the early stages of Cosmogony. Supergravity is an embedding of Einstein's theory in a broader framework, with two main aims: finiteness of Quantum Gravity and (Super) unification with other interactions. This is Highest Energy Physics, i.e. at energies of the order of the Planck mass (10^{19} GeV), 4-5 orders of magnitude beyond GUT physics. It should dominate the creation of matter by the gravitational field, and the emergence of macroscopic asymmetries such as those studied by GUTs. Indeed, for all we know, it may even altogether

---

* Mortimer and Raymond Sackler Institute of Advanced Studies, TAU

** Wolfson Chair Extraordinary, Department of Physics and Astronomy, TAU

*** Supported in part by the U.S. - Israel Binational Science Foundation

† Center for Particle Theory, UT

†† also Center for Theoretical Physics, UT

††† Supported in part by the U.S. DOE under Grant EY-76-S-05-3992

supersede the GUTs or absorb them, rather than just supplementing them above 10^{15} GeV.

2. Progress and Key Results

The key achievements to date have consisted of the following items:

(a) 2-loop finiteness of Quantum Gravity, including the presence of matter - provided that this matter, together with the gravitational field, form an appropriate "extended" supermultiplet.

(b) 3-loop super-renormalizability of a "simple theory" (N=4 Super-Yang Mills), resembling N=8 Extended Supergravity.

(c) vanishing of geometrical counter-terms for n loops, $n < N$.

(d) positive-definiteness of the Gravitational Hamiltonian.

(e) the removal of the Johnson-Sudarshan and other inconsistencies, which used to plague spin 3/2 theories.

All of these developments appear to augur well for a future finite Quantum Gravity. Indeed, they represent great advances since 1974, i.e. since the pioneer results of Veltman and 't Hooft in covariant quantization. In addition, Einstein's Unification Program has made tremendous progress. His dream of Gravity-Electromagnetism Unification has indeed been realized "better" than in any of his own attempts or those of his contemporaries. It is in this context, for instance, that item (e) has been obtained: e.g. a removal of the acausal (faster than light) propagation (the "Velo-Zwanziger paradox") "traditionally" connected with charged spin 3/2 fields. This is important, but does not represent the exalted aims which motivated Einstein in his Search. Our target has changed, and we do not consider that we have reached the goal by writing down a consistent and non-trivial Unification of Gravity with Electromagnetism ("N=2 Extended Supergravity"). Instead we have pursued the program all the way to N=8, the widest Superunification allowed by field-theory as we know it; it is also a particularly promising framework for a finite theory.

Almost every step from N=1 to N=8 Supergravity has brought insights, and I shall devote much of this review to the N=8 theory, including mention of attempts to provide it with a phenomenological interpretation.

3. Supersymmetry

The Super-Poincaré algebra $\hat{p}$ was introduced[1)] by Golfand and Likhtman in 1971[†]. Wess and Zumino worked out[2)] in 1973 the transformations defining the Super conformal group $\hat{C} \supset \hat{P}$. Volkov and Soroka[3)] and Salam and Strathdee[4)] introduced superspace, the factor manifold $\hat{S} = \hat{P}/L$, where L is the Lorentz group $SL(2\mathbb{C})$. Corwin, Ne'eman and Sternberg[5)] clarified the structure of a Lie super-algebra (or graded Lie algebra, GLA) and defined the Extended Superconformal group $\hat{C}_N$ and its subgroup $\hat{P}_N$, known as Extended Supersymmetry. Haag, Lopuszanski and Sohnius[6)] showed that $\hat{P}_N$ and $\hat{C}_N = SU(2, 2/N)$ but[7,8)] $MSU(2, 2/4) = SU(2, 2/4)/I = \hat{C}_4$, I denoting the identity provide the only possibility to unify internal symmetries with those of space-time.

$\hat{C}_N$ is $\mathbb{Z}$ graded, with the grading corresponding to a doubling of the dimension (i.e. of the bracket-eigenvalue of Δ, the dilation generator).

$$\begin{array}{ccccccccc}
\cdots\ c_{-3} & c_{-2} & c_{-1} & c_0 & c_1 & c_2 & c_3\ \cdots \\
\cdots\ 0 & \Xi_\mu & \Psi^a_\alpha & \Sigma_{\mu\nu} & T^a_\alpha & \Pi_\mu & 0\ \cdots \\
 & & & \Delta & & & \\
 & & & \Omega_{ab} & & & \\
 & & & \Theta_{ab} & & &
\end{array} \qquad (3.1)$$

$\mu,\nu = 0,1,\ldots 3$; $a,b = 1,2,\ldots N$ ⌞- - displays $\hat{p} \subset \hat{c}_N$

† we use capital letters for groups, lower case type for algebras, gothic for manifolds, and a caret for "super", or "graded" in the sense of Ref. 5.

$\ell := \{\Sigma_{\mu\nu}\} = s\ell(2,\mathbb{C})$ is the Lorentz algebra, π_μ are the translations, Ξ_μ the special conformal transformations; $r := \{\Omega_{ab}\} = so(N,\mathbb{R})$ generates an orthogonal group; $\{\Theta_{ab}\} = u(N)/r$ is an odd-parity set of generators forming (together with the Ω_{ab}) the algebra of $u(N)$ (except for the case N=4, where it generates SU(4)). For N=1, there is no Ω_{ab}, but Θ_{00} exists as $\Theta = \gamma_5$, the chiral transformations. The odd-grade generators obey the brackets,

$$\{T^a_\alpha\ ,\ T^b_\beta\} = -\frac{1}{2}\,\delta^{ab}\,(\gamma^\mu C)_{\alpha\beta}\,\pi_\mu$$

$$\{\Psi^a_\alpha\ ,\ \Psi^b_\rho\} = \frac{1}{2}\,\delta^{ab}\,(\gamma^\mu C)\,\Xi_\mu$$

$$\{T^a_\alpha\ ,\ \Psi^b_\beta\} = -\frac{1}{2}\,\delta^{ab}\,C_{\alpha\beta}\,\Delta + \delta^{ab}\,(\sigma^{\mu\nu}C)_{\alpha\beta}\,\Sigma_{\mu\nu} + C_{\alpha\beta}\,\Omega^{ab} + i(\gamma_5 C)_{\alpha\beta}\,\Theta^{ab} \tag{3.2}$$

We refer the reader to references[5,6,9)] for the remaining brackets, listing here only those that will be needed for the sequel,

$$[T^a_\alpha\ ,\ \Sigma_{\mu\nu}] = (\sigma_{\mu\nu})^\beta_\alpha\,T^a_\beta$$

$$[T^a_\alpha\ ,\ \Omega_{bc}] = \delta^a_b\,T^b_\alpha - \delta^a_c\,T^c_\alpha$$

$$[T^a_\alpha\ ,\ \Pi_\mu] = 0$$

$$[\Xi_\mu\ ,\ \Pi_\nu] = -2(\eta_{\mu\nu}\Delta + \Sigma_{\mu\nu}) \tag{3.3}$$

Note that T^a_α and Ψ^a_α are "real" Majorana spinors, existing only in space-time dimensionalities which allow the γ^μ to take on pure imaginary values. This happens in 2,3,4 modulo 8 dimensions[10)]. In these dimensionalities $d = 4+N$, an $so(1, N+3)$ spinor has $2^{[(N+3)/2]}$ components, where [ ] denotes the integer part. Such spinors in 4+N dimensions reduce to $2^{[(N-1)/2]}$ real Majorana space-time 4-spinors.

Returning to (3.1), Superspace is the factor manifold

$$\hat{S}_N = \hat{P}_N/L \times R$$

where L is the Lorentz group manifold and R that of the $SO(N,\mathbb{R})$ internal orthogonal symmetry.

4. Gauging on a Soft Group Manifold

In Supergravity[11,12] superspace $\hat{S}_N(\chi^\mu, \Theta^{\alpha a})$ is curved and all brackets (3.3) and (3.4) etc. may acquire additional terms on the right hand side (:= and =: represent a definition)

$$[x_{curved}, y_{curved}] := [x,y] + \sum_A R^A_{xy}\, z_A \tag{4.1}$$

where R^A_{xy} is thereby a "curvature" by definition. As I showed with T. Regge[13)], and developed with J. Thierry-Mieg[14)] (see also Ref. 15), we are dealing here with a "Soft-Group-Manifold" (SGM) undergoing a fibration as a result of the equations of motion. The SGM is a triplet $\mathcal{G}:=(G,\mathcal{G},\rho)$ where G is the non-internal group (such as P, $\hat{P}$, $\hat{P}_N$ etc...), $\mathcal{G}$ is a differentiable manifold of the same dimensionality "m", and ρ^A is a set of frames over $\mathcal{G}$, behaving under G as its adjoint representation. Explicitly this is a one-form

$$\rho^A = \rho^A_M\, d\,\zeta^M \quad , \tag{4.2}$$

where ζ^M is the coordinate over $\mathcal{G}$. Picking an action

$$S = \int_{\mathcal{R}^d} R^A \wedge T_A \quad , \tag{4.3}$$

$\mathcal{R}^d$ is Riemannian d-dim, spacetime, $R^A := R^A_{MN}\, d\zeta^M \wedge d\zeta^N$ and T_A are respectively 2-forms and $(m-2)$ forms. The T_A are chosen so as to break G and preserve local F-symmetry, $F \subset G$. For instance, $F := ((\Sigma_{\mu\nu}, \Omega_{ab})) = s\ell(2,\mathbb{C}) \otimes so(N)$ in $\hat{P}_N$. The equations of motion enforce in $\mathcal{G}$ a fibration ($\overset{0}{=}$ denotes equality "on mass shell")

$$\overset{F}{\mathcal{G}} :\overset{0}{=} \mathcal{G} \;, \quad \overset{F}{\mathcal{G}}\,(\overset{F}{\mathcal{G}},\, F,\, \mathcal{G}/F,\, \pi,\, \cdot) \tag{4.4}$$

and R^A or T_A become constrained as forms to be "horizontal" $H \overset{0}{=} \mathcal{G}/F$,

$$F :\overset{\Omega}{=} \mathcal{F}$$

which is in fact superspace $\hat{S}_N$. The fibration implies ({F} represent the range of indices of F)

$$B \text{ or } C \in \{F\} \;, \rightarrow \; R^A_{\;BC} \overset{\Omega}{=} 0 \quad . \tag{4.5}$$

Besides the above result, we also find that the generalized torsion vanishes

$$R^A \overset{\Omega}{=} 0, \quad \forall \{A\} \in G/F \quad . \tag{4.6}$$

A sufficient condition[13,14] for G to fibrate, for the relevant groups[14]
G is that they be weakly-reducible,

$$\left.\begin{aligned} & g = f \oplus h \\ & [f,f] \subset f \\ & [f,h] \subset h \end{aligned}\right\} \tag{4.7}$$

and in addition have a symmetric decomposition

$$[h,h] \subset f \quad . \tag{4.8}$$

The latter condition is not fulfilled by the (3.3) commutator $\{T\ T\} \subset \Pi$, since both T and Π are in h, for $F = ((\Sigma))$, $g = \hat{p}$. Supergravity can be regarded either as an incomplete fibration, or, for $N = 1$, the action can be represented by the sum of two fibrated G^+ and G^-, where f is chosen as[13,14]

$$f^{\pm} = \ell \oplus ((T^{\pm})) \tag{4.9}$$

and $T^{\pm}$ are the Left and Right Chiral components. As a result,

$$\left.\begin{aligned} & R^i \overset{\Omega}{=} 0 && i,j = 0,1..3 \\ & R^A_{\;[ij]B}\, \rho^{[ij]} \wedge \rho^B \overset{\Omega}{=} 0 && a,b = 1...N \\ & R^A_{\;[ab]B}\, \rho^{[ab]} \wedge \rho^B \overset{\Omega}{=} 0 && [ij] \in \{s\ell(2\mathbb{C})\} \\ & R^i_{\;\alpha a,\beta b} \overset{\Omega}{=} 0 && [ab] \in \{so(N)\} \end{aligned}\right\} \tag{4.9}$$

and α and β are Majorana spin indices.

The "curved" generators of equation (4.1) are given by an orthonormal basis $\tilde{a}$ of "vector fields" (mathematical appelation) in the Soft Group Manifold G

$$\rho^A \lrcorner \tilde{a}_B = \delta^A_B \quad . \tag{4.10}$$

The $\tilde{a}_i$ and $\tilde{a}_\alpha$ are in fact the corresponding "covariant derivatives" and when locally gauged, correspond to the "Lie derivative" by an anholonomic vector field $\tilde{\varepsilon}^B$

$$\delta\rho^A = L_{\tilde{\varepsilon}} \rho^A = D\varepsilon^A + \varepsilon^B \wedge \rho^C R^A{}_{BC} \tag{4.11}$$

"Local supersymmetry", the transformation generating Supergravity, indeed corresponds to $\bar{\varepsilon}_\alpha{}^a$ in (4.11). It is an "anholonomized" general coordinate transformation, and we shall derive the variations for the various gauge fields. However, we can already note that the last equation in (4.9) guarantees

$$\{\tilde{T}^a_\alpha, \tilde{T}^b_\beta\} \overset{o}{=} -\frac{1}{2}\delta^{ab} (\gamma^\mu \tilde{\pi}_\mu)$$

which, when applied to a state at rest, yields on mass shell

$$\{\tilde{T}^a_\alpha, T^b_\beta\} |\vec{p} = 0 > \overset{o}{=} \frac{1}{2} \delta^{ab} \tilde{\Pi}_0 | \vec{p} = 0 > \tag{4.12}$$

where $\tilde{\Pi}_0$ is the Hamiltonian (or the supercovariant derivative). This can be shown to guarantee[16)] positive-definiteness of the gravitational Hamiltonian, even after a gradual extinction of the "super" contributions. However, (4.12) is written in superspace $\hat{S}(x^\mu, \Theta^{\alpha a})$ and when we restrict to spacetime, the algebra does not close[17)].

Note that Ω_{ab} in (3.2) disconnects from the rest of $\tilde{P}_N$. It acts on the T^a_α, which behave under it as an SO(N) vector, but it is not reproduced by a commutator or anticommutator of $\tilde{P}_N/SO(N)$. As a result, it is not gauged locally with the factor group $\tilde{P}_N/SO(N)$.

An alternative approach involves working on $\hat{S}_N$ off-mass-shell as well as on-mass-shell. The action cannot correspond to (4.3) and has generally been taken to be identical with the measure over superspace[18], with appropriate constraints[19-20]. This "direct" superspace (DS) method (or an equivalent set) has been applied to N=1 and higher N. For N=1, the $\bar{\varepsilon}^{\alpha}$ variations have been closed off-shell by identifying the appropriate curvature components (4.1) with new superfields (fields in χ, Θ).

For quantum Gravity to be finite (it cannot be just renormalizable, due to the dimensional Newtonian coupling κ) within a renormalizable quantum Supergravity written over $\hat{S}_N$ (and not G), we have to prove that no counter-terms can appear at any order.

5. Extended Supergravity Representations

Note the δ^{ab} on the right-hand side of (4.12); the coupling of internal and external degrees of freedom is non-trivial. This is why Supergravity is an advance over the Kaluza-Klein solution. Over there, the addition of a (compactified) dimension produces an elegant derivation of the Einstein-Maxwell theory, but with no new mathematical constraints. It is just a juxtaposition, (not a coupling) of the two interactions. We had generalized it[21,22] to $su(3) \subset su(4) = \overline{so}(6)$ by embedding $R^4 \subset M^{10}$, in an early attempt to renew the search for Unification, hypothetizing that "flavor" symmetries emerge through the short-range character of the Strong Interactions. In Extended Supergravity, however, equation (4.12) couples the internal degrees of freedom to the external.

Particle states can be defined in a local frame. We then have the representations of $\hat{P}_N$, given by the little group. For (3.3) acting on massive states[23], we have (4.12) which defines a 4N-dimensional Clifford algebra. For N=1 the 4 states are given by $(s, s+\frac{1}{2}, s-\frac{1}{2}, s')$, where s is some basic helicity $\Sigma_{12} = f_3 = s$. For massless states (or for $\hat{C}_N$), we use[5] $p^{\pm} = p_0 \pm p_3$. The system reduces into $p^+ \neq 0$, $p^- = 0$ and vice versa. At $p^{\pm} \to \infty$, this is identical with R/L chiralities.

The massless representations of Supergravity,[24] for N=1 are 2-dimensional (helicities f_3 = s, s-½) but CPT invariance requires a doubling, adjoining (-s, -s + ½). For N, the number of states at each helicity $j_3 = j_3^{max} - k/2$ (using lowering operators $T^-_{\alpha a}$) in an irreducible representation is given by the binomial coefficient $\binom{N}{k}$, as befits Clifford or Grassmann algebras in which the anticommutator is a number (or zero), and only the antisymmetrized combination contributes to the enveloping algebra. Symbolically, the representation comprises in terms of Young tableaux,

$$|j_3^{max}\rangle \oplus \square^{a}_{\Delta j_3 = -\frac{1}{2}} |j_3^{max}\rangle \oplus \begin{array}{|c|}\hline \\ \hline \\ \hline \end{array} |j_3^{max}\rangle \oplus \dots \oplus \begin{array}{|c|}\hline \\ \hline \\ \hline \\ \hline \\ \hline \end{array}\begin{array}{l} 1 \\ 2 \\ \vdots \\ N \end{array} |j_3^{max}\rangle \tag{5.1}$$

For N=1 we get helicities (2; 3/2) ⊕ (-3/2; -2), the spin 2 graviton and spin 3/2 "gravitino". For N=2 we have (2; 2 x 3/2; 1) ⊕ (-1; 2 x (-3/2); -2). This is the realization of Einstein's aim: a graviton, the electromagnetic field with spin 1, and a charged spin 3/2 gravitino. The system continues to grow all the way to N=7,

(2; 7x3/2; 21x1; 35x½; 35x0; 21x(-½); 7x(-1); -3/2; -)

⊕ (-; 3/2; 7x1; 21x½; 35x0; 35x(-½); 21x(-1); 7x(-3/2); -2)

(5.2)

whereas for N=8 we have one irreducible representation, though with the same particle-state composition as N=7

(2; 8x3/2; 28x1; 56x½; 70x0; 56x(-½); 28x(-1); 8x(-3/2); 2) .

(5.3)

Note that though the spin 1 states do fit the adjoint representation of SO(N), they do not correspond to the gauge fields of local SO(N), which does not become gauged in $\hat{P}_N$. In fact, all fields in these representations correspond to a "2nd order" formalism and are not the direct (1st order) connections. They are related to them by equations of motion, constraints or Bianchi identities, depending on the formalism. This is just as in Gravity, where the vanishing of the torsion $R^i = 0$ is used to solve $\rho_\mu^{[ij]}$ (the Lorentz connection) in terms of ρ_μ^{i} (the tetrad) and reproduce the Christoffel formula. The $j = 3/2$ field in the representation is related to $\rho_\mu^{\alpha a}$ in a complicated way, since $j = 2$ stands for $g_{\mu\nu}$ rather than ρ_μ^{i} or $\rho_\mu^{[ij]}$

The irreducibility of N=8 is typical of systems in which $f_3^{max} = N/4$. One such theory is the "special" N=4 Yang-Mills supersymmetry[25,26] with

$$(1;\ 4\times\tfrac{1}{2};\ 6\times 0;\ 4\times(-\tfrac{1}{2});\ -1) \tag{5.4}$$

This theory has recently[27-28] been shown to have vanishing coupling constant renormalization[29] $\beta = 0$ to 3 loops,

$$\lambda \frac{dg(\lambda)}{\lambda} = \beta(g(\lambda)) \tag{5.5}$$

These irreducible particle representations of N=extended supersymmetry also have in common the fact that their particle content is the same as that of the reducible N-1 case.

The $\hat{P}_N$ representations are also representations[30] of su(1/N), as can be checked by comparing with Refs. (31)-(33). Since the particles are massless, they obey the superconformal algebra su(2, 2/N). The latter's representations are induced by the "little group".

For the irreducible $N = 4f_3^{max}$ cases, we can extend SU(1/N) by the discrete CPT, since the representations are also eigen-representations of CPT. We may thus define a super-unitary parity[30] η_S, taking for instance the CPT parity of the $f_3 = 0$ symmetric combination of the "middle" states.

The $\beta(N=4) = 0|_{n \leq 3}$ result may augur well for a finite theory. It is possible that the theory is indeed finite to all orders, and that the same be true of N=8 supergravity!

6. Supergravity with N=1, 2

The Lagrangian (4.3) for N=1 Supergravity can be written on spacetime as a 4-form

$$
\begin{aligned}
L &= R^{ij}\, T_{ij} + \bar{T}_\alpha\, R^\alpha \\
T_{ij} &= \varepsilon_{ijkl}\, e^k \wedge e^l \\
T^{\bar{\alpha}} &= 4i\, \bar{\psi}\, \gamma^j\, e^j\, \gamma_5 \\
R^{ij} &= d\omega^{ij} + \omega^{ik} \wedge \omega^{kj} \\
R^\alpha &= d\,\psi^\alpha + \frac{1}{2}\left(\omega^{ij}\, \sigma_{ij}\, \psi\right) = \left(D^L\, \psi\right)^\alpha
\end{aligned}
\qquad (6.1)
$$

where e^i, ω^{ij} and ψ^α are the R^4 restrictions of the connections[13,14] ρ, $\rho^{[ij]}$ and ρ^α. The two terms in L are the Einstein and the Rarita-Schwinger Lagrangians.

For N=2, writing directly in 2nd order formalism, we have[35,36]

$$
\begin{aligned}
L = &- \frac{1}{2\kappa^2}\, \Theta\, R - \frac{1}{2}\, \varepsilon^{\mu\nu\rho\sigma}\, \psi^a_\mu\, \gamma_5\, \gamma_\nu\, D^L_\rho\, \psi^a_\sigma \\
&- \frac{1}{4}\, e\, g^{\mu\rho}\, g^{\nu\sigma}\, F_{\mu\nu}\, F_{\rho\sigma} \\
&+ \frac{\kappa}{\sqrt{2}}\, \bar{\psi}^1_\mu \left[e F^{\mu\nu} + \frac{1}{2}\, \gamma_5\, {}^*F^{\mu\nu}\right] \psi^2_\nu \\
&+ \text{quartic terms in } \psi^a_\mu \\
&(e \text{ is } \det e^i)
\end{aligned}
\qquad (6.2)
$$

The index $a = 1,2$. We have written the result explicitly, so as to display this solution to Einstein's quest. It involves a specific magnetic coupling of A_μ to ψ^a_μ, and it is the constraints of "local supersymmetry" which do away with negative probabilities[37] and acausal propagation[38]. Moreover, the theory is <u>2-loop finite on mass shell</u> whereas ordinary Einstein-Maxwell theory fails at the 1st loop[39].

It is also possible to impose local SO(2) gauge invariance[40]. This produces a different Lagrangian with a cosmological term of Planck size. The particle interpretation is therefore unclear, so is quantization. On the other hand, it is possible that such a geometry could become the correct quantum solution, if indeed spacetime could be taken as quantized geometrically (the Wheeler "foam") as recently discussed[41].

7. The N=8 Extended Supergravity

As we had originally observed[24], the system of Extended Supergravity ends at N=8, if there is no new input capable of dealing with the emergence of several gravitons. At the same time, there is a hope for a finite N=8 theory, deriving from the present success in N=4 Yang-Mills.

Writing Lagrangians directly for larger N becomes difficult after N=3. With a hint from Nahm[42] and using the adaptation[43] of our [21-23] Kaluza-Klein approach to Supersymmetry, Cremmer and Julia[44,45] have written the N=8 case by first going through 11-dimensional N=1 Supergravity. The latter is given by

$$\kappa^2 \mathcal{L}^{d=11} = -\frac{e}{4} R - \frac{e}{48} F_{MNPQ} F^{MNPQ} - i\frac{e}{2}\bar{\psi}_M \Gamma^{MNP} D_N \left(\frac{\omega+\hat{\omega}}{2}\right)\psi_P$$

$$+ \frac{e}{192}\bar{\psi}_M X^{MNWXYZ}\psi_N (F_{WXYZ} + \hat{F}_{WXYZ})$$

$$+ \frac{2}{144}\varepsilon^{M_1M_2\ldots M_{11}} F_{M_1M_2M_3M_4} F_{M_5M_6M_7M_8} A_{M_9M_{10}M_{11}} \qquad (7.1)$$

all (holononic) vector indices MN...Z go from 0 to 10. As we see, besides the 11-dim. graviton g_{MN} (or e_M^A plus ω_M^{AB} with only one kinetic energy term) we have as gauge fields a 32-component Majorana SO(10,1) spinor (anholonomic m index), with an additional 11-vector (holononic) index $\psi^{m=1\ldots32}_{M=0\ldots10}$; also, and a three-index totally antisymmetric gauge field of the Nambu or Kalb-Ramond type, corresponding to a tensor gauge function α_{PQ}. Such a field has zero physical content in spacetime M^4, but does exist in M^{11}. In addition,

$$\left.\begin{aligned}
&\Gamma^{MN\ldots Z} = \Gamma^{[M}\,\Gamma^{N}\,\ldots\,\Gamma^{Z]} \\
&F_{MNPQ} = 4\,\partial_{[M}\,A_{NPQ]} \\
&X^{MNWXYZ} = \Gamma^{MNWXYZ} + 12\,g^{M[W}\,\Gamma^{XY}\,g^{Z]N} \\
&\hat{\omega}_M{}^{AB} = \omega_M{}^{AB} + \frac{i}{4}\,\bar{\psi}_N\,\Gamma_M{}^{ABNP}\,\psi_P \\
&\hat{F}_{MNPQ} = F_{MNPQ} - 3\,\bar{\psi}_{[M}\,\Gamma_{NP}\,\psi_{P]}
\end{aligned}\right\} \qquad (7.2)$$

Reduction to N=8, d=4, Supergravity proceeds[46)] through the assumption of non-dependence of the fields on the variables x^{3+a}, $1 \le a \le 7$. The action is then integrated over d^4x only. The last seven dimensions are compactified to circles, whose radii are then taken to zero.

To recover the fields corresponding to the particles in (5.3) one has to take various functions of those in (7.1). First, the local $\overline{SO}(10, 1)$ symmetry of the frame indices (gauged by $\omega_M{}^{AB}$) is used to fix a gauge in which the spacetime tetrad will have no holonomic index M of value larger than 3, i.e. 28 conditions $e_n{}^i = 0$,

$$e_M{}^A = \begin{pmatrix} e_\mu{}^i & B_\mu{}^a \\ 0 & s_n{}^a \end{pmatrix} \qquad \begin{matrix} \mu = 0\ldots3 & i = 0\ldots3 \\ n = 4\ldots10 & a = 4\ldots10 \end{matrix} \qquad (7.3)$$

so that $e_M{}^A$ yields a vector-meson B_μ^a behaving as an SO(7) 7-vector that still gauges 7 internal ζ^a circular "translations". Aside from Einstein-covariance (gauged by $e_\mu{}^i$) the diffeomorphisms have also left a residual global GL(7, R) as an internal symmetry, together with a multiplet $s_n{}^a$, equivalent to a GL(7, R) symmetric $g_{mn}\underline{28}$ tensor of spacetime scalar fields. The local $\overline{SO}(10, 1)$ of the frames has now been replaced by local Lorentz invariance $\overline{SO}(3, 1)$ gauged by $\omega_\mu{}^{ij}$ and a local $\overline{SO}(7)$ gauged by $\omega_M{}^{ab}$. All of the above gauges correspond to the Einstein-like-term in (7.1). The $\psi_M{}^m$ field yields 8 Rarita-Schwinger gravitinos $\psi_\mu^{\alpha u}$ (u = 1...8) supergauging the above-mentioned internal $\overline{SO}(7)$ as in (5.2) but capable of gauging an $\overline{SO}(8)$ as in (5.3); and 56 spin 1/2 fields $\psi_a{}^{\alpha u}$. The A_{MNP} which gauged 55 U(1) groups now reduce to a non-physical auxiliary $A_{[\mu\nu\rho]}$, 7 scalars $A_{[\mu\nu]m}$, 21 vector $A_{\mu[mn]}$ and 35 scalar $A_{[mnp]}$. Appropriate duals or functions of all of these are then regrouped to fit (7.1)

$$(J = 2 : g_{\mu\nu};\ J = \frac{3}{2} : \psi_\mu{}^{\alpha u};\ J = 1 : A_{\mu[mn]} \to B_\mu{}^{mn},\ B_\mu{}^a \text{ or } B_\mu^m ;$$

$$J = \frac{1}{2} : \psi_m{}^{\alpha u} \to \psi^\alpha_{[uvw]};\ J = 0 : s_n{}^a \text{ or } g_{mn},\ A_{[\mu\nu]m} \to \phi^m \ ,$$

$$A^{(-)}_{[mnp]} \to {}^{\star}A^{[mnpq]} \tag{7.4}$$

The Lagrangian then involves $\partial_\mu\phi^m$ and $G_{\mu\nu}{}^{mn} = 2\partial_{[\mu} B_{\nu]}{}^{mn}$, with $B_\mu{}^{mn}$ gauging $[U(1)]^{\otimes 21}$ and $B_\mu{}^a$ gauging $[O(2)]^{\otimes 7} \sim [U(1)]^{\otimes 7}$. In addition, we have, as discussed above, a gravitational-type $\overline{SO}(7)$ local gauge and GL(7, R) global symmetry. These internal symmetries can now be prolongated.

$$\overline{SO}(7)_{local} \times GL(7, R)_{global} \to \overline{SO}(8)_{local} \times SL(8, R)_{global} \tag{7.5}$$

first by completing $\overline{SO}(8)$ multiplets everywhere, i.e. going in (7.4) from (5.2) to (5.3). The Γ^i can be adjoined to the Γ^{ij} to yield the spinorial 8-dimensional matrix representation of $\overline{SO}(8)$, thus making the $\psi_\mu{}^{\alpha u}$ into an $\overline{SO}(8)$ spinor. For J = 1, $B_\mu{}^a$ becomes $B_\mu{}^{m11}$, using an inverse internal tetrad and defining a formal 11th spatial internal dimension. Similarly, ϕ^m becomes ϕ^{m11} and is adjoined to the (28) g_{mn} as $g_{(m11)}$ after holonomisation of the index a into m by an internal inverse tetrad $s_a{}^m$, to make an SO(8) symmetric traceless <u>35</u>. With appropriate Jacobians etc. the inverse $g^{m'n'}$ (m',n' = 4...11, i.e. 8 values) can be replaced by

$$S^{m'n'} = E^{-3/4} \begin{pmatrix} Eg^{mn} - \phi^m \phi^n \ , & \phi^{n11} \\ \phi^{m11} & -1 \end{pmatrix}, \quad \begin{matrix} E = -\det g_{mn} \\ \det S = 1 \end{matrix} \tag{7.6}$$

The GL(7, R) global analog of Einstein covariance has been replaced by SL(8, R). We can now return to factorized "tetrads" $s_{m'n'} = t_{m'}{}^{a'} \eta_{a'b'} t_{n'}{}^{b'}$ and enlarge the local $\overline{SO}(7)$ of the frames to local $\overline{SO}(8)$ (the dashed indices take 8 values). This $\overline{SO}(8)$ group could be gauged through the scalar $s_{m'n'}$, Einstein-wise, except for the action on spinors, which requires the $t_{m'}{}^{a'}$. The $s_{m'n'}$ can be taken as the Goldstone fields of SL(8, R)/SO(8) to realize this global invariance, even with non-compact SL(8, R). This corresponds to a special "gauged" σ-model for a group G, in which the compact subgroup $H \subset G$ is locally gauged. The H currents themselves supply the non-propagating gauge field, and cancel the (negative) contribution of the compact part. Effectively, only the G/H Goldstone contribution survives, with positive energy. Another such prolongation occurs through similar use of an auxiliary $J=1$ field, to gauge an anholonomic SU(8) group. The field is given by the $v^{-1}\partial_\mu v$ currents of all 70 scalar fields v, in a rather complicated manner which we shall summarize herewith.

The global (holonomic) SL(8, R) is prolongated to $E(7)_7$, a non-compact form of E(7), with SU(8)/Z(2) as compact subgroup. This E(7) is an on-shell symmetry because it is realized through the <u>duals</u> of the $J=1$ field strengths ($B_\nu{}^m$ becomes $B_\nu{}^{m11}$)

$$G_{\mu\nu}{}^{m'n'} = 2\,\partial_{[\mu} B_{\nu]}{}^{m'n'} \tag{7.7}$$

yielding

$$H^{\mu\nu}{}_{m'n'} = \frac{4}{e} \frac{\delta L}{\delta G_{\mu\nu}{}^{m'n'}} \tag{7.8}$$

Together, the $G_{\mu\nu}{}^{m'n'}$ and the $H^{\mu\nu}{}_{m'n'}$ form a <u>56</u>, which is defined as the $\underline{56}$ of $E(7)_7 \supset SL(8, R)$,

$$F_{\mu\nu} = \begin{pmatrix} G_{\mu\nu}^{\ m'n'} \\ \\ H_{\mu\nu\ m''n''} \end{pmatrix} \tag{7.9}$$

with $F_{\mu\nu}$ satisfying a constraint, since $G_{\mu\nu}$ and $H_{\mu\nu}$ are not independent. New "internal quasi-tetrads" perform the transition between the $(m'n', m''n'')$ $\underline{56}$ indices of the global (holonomic) $E(7)_7$, and the $(a'b', a''b'')$ $\underline{56} = \underline{28} + \underline{28}$ of its $SU(8)/Z(2)$ compact subgroup, with

$$SU(8)/Z(2) \cap SL(8, R) = SO(8) \quad .$$

This $SO(8)$ is a holonomic global symmetry, as a subgroup of $E(7)_7 \supset SL(8, R)$, $SL(8, R)$ prolongating $GL(7, R)$. There is, on the other hand an anholonomic (Einstein-local) $\overline{SO}(8)$ which prolongated the original (internal) $\overline{SO}(7)$ of the spinors etc. The final N=8 SO(8) linear (an global) internal symmetry contained in $\hat{P}_8$ is a global subgroup of the diagonal sum of the two.

Returning to the $G_{\mu\nu}^{\ m'n'}$ and $H_{\mu\nu}^{\ m''n''}$, we are interested in replacing m', n' and m'', n'' by a', b' and a'', b'' in order to achieve invariant couplings to other particles, thus involving double-use of antisymmetrical $t_{[m'}^{\ [a'} t_{n']}^{\ b']}$ containing as we saw $\underline{7} + \underline{28} = \underline{35}$ spinless mesons.

These then provide 28×28 quasi-tetrads for $G_{\mu\nu}^{\ m'n'}$ or $H_{\mu\nu}^{\ m''n''}$. The other $\underline{35}$ of 0^- mesons is represented by the 35 duals of $A_{[mnp]} \rightarrow {}^*A^{[mnpq]}$. With some re-indexing, these are made to fit into the off-diagonal 28×28 squares of the above-mentioned 56×56 system of internal quasi-tetrads, with the $t_{[m'}^{\ [a'} t_{n']}^{\ b']}$ in the diagonal squares. $E(7)_7$ now provides an invariant coupling of all $J = 1,0$. The group has 133 parameters, and the same gauged σ-model method can be used to write an invariant mesonic Lagrangian. The non-compactness is again taken care of by extending the $SU(8)/Z(2)$ subgroup as a local gauge, with an auxiliary 63 gauge-field constructed from currents. The physical degrees of freedom indeed correspond to 133-63 = 70 mesons.

The fermions have the correct dimensionalities for SU(8), and an (anholonomic) prolongation $\overline{SO}(8)_{local} \rightarrow SU(8)_{local}$ can be performed, using the same type of composite (non-propagating) gauge field, with the help of 63 - 28 = 35 J = 0 mesons. $E(7)_7$ is holonomic, and does not act here.

Once this construction is reached, it is of course possible to write down the final result directly, forgetting the auxiliary d = 11 Lagrangian. Extended N=8 Supergravity, which could only be approximated in a direct approach[47], is thus now known to allow an $E(7)_7 \supset SL(8, R)$ holonomic global symmetry, another local anholonomic $SU(8) \supset \overline{SO}(8)$, and of course $\hat{P}_8 \supset \hat{P}_{local} \times SO(8)$. There is a particular global SO(8) in the intersection of all three symmetries. Local symmetries are not of the Yang-Mills type. In addition, the theory possesses an SU(1/8) residual super conformal symmetry, manifest in the representation of the physical particles, and group-extendable by the discrete CPT parity.

Spontaneous breakdown and mass generation have been constructed[48], this time using as an auxiliary step a reduction $\hat{P}_1(M^{11}) \supset \hat{P}(M^5)$ with a resulting global $E(6)_6 \supset Sp(8)$ and a corresponding $Sp(8)_{local}$ symmetry. Sp(8) also appears in a related off-shell supersymmetric version[49] of $\hat{P}_8$, as the manifest global invariance of the theory.

Returning to the last paragraph of section 6, note that N=1 gravity with a cosmological constant (the Einstein universe of 1917)

$$S = -\frac{1}{2^K}\int d^4 x \sqrt{g}\,(R - 2\Lambda) \tag{7.10}$$

is not one-loop finite[50]. The counter term (ε = n-4 is the dimensional regularization variable)

$$\Delta\tilde{S} = -\frac{1}{\varepsilon}(A\chi + B\delta)$$

$$X := \frac{1}{32\pi^2}\int d^4\underline{x}\,\sqrt{g}\;{}^*R_{\mu\nu\rho\sigma}\;{}^*R^{\mu\nu\rho\sigma}$$

$$\delta := \frac{1}{12\pi^2}\int d^4x\,\sqrt{g}\,\Lambda^2 = -\frac{\kappa^2\Lambda}{12\pi^2}\,S \qquad (7.11)$$

has $A = 106/45$, $B = -87/10$. The same non-finiteness is true of simple supergravity. However, for $N > 4$ theories, $B = 0$, and in addition, the one-loop contribution to β vanishes. Moreover, X takes on integer values, indicative of a new index theorem. Siegel has now pointed out that due to quantum-inequivalence, the particular set of representations with $J = 0$ in (7.4) yields $A = 0$ as well as $B = 0$ in N=8 Extended Supergravity! This implies vanishing of the trace anomaly, since X is the topological counterterm. Duff has also discovered that even the boundary contributions to the one-loop counterterms, which are normally ignored, also cancel. N=8 supergravity is thus the only known theory in which all four types of counterterms allowed in quantum gravity cancel (at least for one loop): conventional ones such as R^2 and $R_{\mu\nu}R^{\mu\nu}$, topological ones (Euler number), cosmological constant terms as in (7.11) and boundary contributions.

It is <u>conjectured</u> that the appearance of $E(7)_7$ is related to the imposition of an S^7 geometry on the internal compactified system, with a further transition to octonions (S^7 is the 7-sphere).

Much of the more recent work has consisted in an attempt at a phenomenological interpretation. Knowing that SO(8) is too small to contain the conventionally assumed essential components in a GUT, $SU(3)_{color} \times [SU(2) \times U(1)]_{weak,\ electr.}$, the emergence of $SU(8)_{local}$ may provide an opening. Meanwhile, it has become more common to assume that quarks and leptons are not elementary. This is especially indicated[51)] by seriality, i.e. the repetitive "generation" structure (ν_e, e, u, d) (ν_μ, μ, c, s) $(\nu_\tau, \tau, t?, b)$ .... Various models have been suggested heuristically[51,52)]. Dynamically, this might be the result of additional "hidden" color-like symmetries, acting at ranges of $< 10^{-17}$ cm. The subconstituents of quarks and leptons are assumed to be confined within these shorter ranges.

To bridge between phenomenology and the composite-particle spectrum of Extended N=8 Supergravity, Ellis *et al.*[53,54] assume that nature realizes an SU(1/8) multiplet containing the $SU(8)_{LOCAL}$ currents, i.e. the composite 63 of $J=1^-$ mesons. This supermultiplet is constructed as in (5.1), by taking as the basic state an SU(8) 8^* with $j_3^{min} = -3/2$ and operating with a (raising) generator 8, $\Delta j_3 = 1/2$,

d(SU(8)):	8^*	1	8	28	56	70	56^*	28^*	8^*
	-	63	216	420	504	378^*	168^*	36^*	-
j_3 :	-3/2	-1	-½	0	½	1	3/2	2	5/2

(7.12)

It is assumed that this multiplet materializes as physical propagating particles, through some unknown dynamics. Our phenomenological "fundamental" fermions (quarks and leptons) and the Higgs mesons required in GUTs are just the low energy piece of this supermultiplet (in this approach). The true fundamental fields are then those appearing with the graviton in N=8 11-dimensional supergravity.

An attempt at a naive implementation of this program is rocked by anomalies in the currents, preventing the construction of a renormalizable theory. A more recent version evades these difficulties by conjecturing that the unknown dynamics cause the symmetries of Extended Supergravity to break down at or near the Planck mass. The SU(5) GUT is then assumed to emerge as an effective lower energy renormalizable theory. Its renormalizability itself has something to do with the lowered energy-scale at which it is realized. Following are the collected sets of SU(5) anomaly-free multiplets contained in (7.12).

For $j = ½$, $8^* + 216^* + 56 + 504$ contain,

$$(45 + 45^*) + 4 \times (24) + 9 \times (10 + 10^*) + 3 \times (5 + 5^*) + 3 \times (5^* + 10) + 9 \times (1) \qquad (7.13)$$

Most of (7.13) is chiral-symmetric ("vector-like"), except for 3 generations (a prediction!) of left-handed 5* + 10. It is further assumed that one set of j=0 5 + 5* and 24 also appear lower than at Planck mass, so as to realize the Georgi-Glashow SU(5) model.

As we see, this phenomenological approach involves much hand-waving. Still it represents a possible way in which Extended Supergravity might father the more conventional GUT.

References

1. Yu. A. Golfand and E.P. Likhtman, JETP Letters 13, 452 (1971).
2. J. Wess and B. Zumino, Nucl. Phys. B70, 39 (1974).
3. D.V. Volkov and V.A. Soroka, JETP Letters 18, 529 (1973).
4. A. Salam and J. Strathdee, Nucl. Phys. B76, 477 (1974).
5. L. Corwin, Y. Ne'eman and S. Sternberg, Rev. Mod. Phys. 47, 573 (1975).
6. R. Haag, J.T. Lopuszanski and M. Sohnius, Nucl. Phys. B88, 257 (1975).
7. P.G.O. Freund and I. Kaplansky, J. Math. Phys. 17, 228 (1976).
8. V.G. Kac, Func. Analysis (USSR) 9, 91 (1975).
9. B. Zumino, in Eighth Texas Symposium on Relativisitic Astrophysics, Ann. NY Acad. Sciences, 302, p. 545-556 (1977).
10. J. Scherk, in Proceedings of Nato Advanced Study Insitute on Gravitation, Cargese, 1978.
11. D.Z. Freedman, P. van Nieuwenhuizen and S. Ferrara, Phys. Rev. D13, 3214 (1976); D14, 912 (1976).
12. S. Deser and B. Zumino, Phys. Lett. 62B, 335 (1976).
13. Y. Ne'eman and T. Regge, Riv. Nuovo Cim. 1, N.5 (1978).
14. J. Thierry-Mieg and Y. Ne'eman, Ann. Phys. 123, 247 (1979). Y. Ne'eman, E. Takasugi and J. Thierry-Mieg, Phys. Rev. D22, 2371 (1980).
15. H. Sokolik, Nuovo Cim., 51A, 339 (1979).
16. M.T. Grisaru, Phys. Lett. 73B, 207 (1978).
17. D.Z. Freedman and P. van Nieuwenhuizen, Phys. Rev. D14, 912 (1976).
18. J. Wess and B. Zumino, Phys. Lett. 74B, 51 (1978).
19. W. Siegel, in Group Theoretical Methods in Physics, Proc. 7th Int. Conf. (Austin 1978) Lect. Notes on Phys. 94, Springer Verlag (Berlin, Heidelberg, NY) p. 487 (1979).

20. L. Brink, M. Gell-Mann, P. Ramond and J.H. Schwarz, Phys. Lett. 74B, 336 (1978).

21. J. Rosen and Y. Ne'eman, Ann. Phys. 31, 391 (1965).

22. Y. Ne'eman, Rev. Mod. Phys. 37, 227 (1965).

23. A. Salam and J. Strathdee, Phys. Rev. D11, 521 (1975).

24. M. Gell-Mann and Y. Ne'eman (unpublished).

25. A. Salam and J. Strathdee, Phys. Lett. 51B, 353 (1974).

26. L. Brink, J.H. Schwarz and J. Scherk, Nucl. Phys. B121, 77 (1977).

27. M. Grisaru, M. Rocek and W. Siegel, Brandeis preprint (1980).

28. L.V. Avdeev, O.V. Tarasov and A.A. Vladimirov, Phys. Lett. 96B, 94 (1980).

29. E.C.G. Stueckelberg and A. Petermann, Helv. Phys. Acta, 26, 499 (1953);
M. Gell-Mann and F. Low, Phys. Rev. 95, 1300 (1954).

30. Y. Ne'eman, Proc. Xth Texas Symp. on Rel. Astrophysics, to be pub. in Ann. N.Y. Acad. Sciences.

31. Y. Ne'eman and St. Sternberg, Proc. Nat. Acad. Sci. 77, 3127 (1980).

32. P.H. Dondi and P.D. Jarvis, Z. Phys. C, 4, 201 (1980).

33. A. Baha Balantekin and I. Bars, Yale preprint YTP 80-86 (1980).

34. B. Morel and J. Thierry-Mieg, Com. Math. Phys. to be pub. and Phys. Letts. 101B, 393 (1981).

35. S. Ferrara, J. Scherk and P. van Nieuwenhuizen, Phys. Rev. Lett. 37, 1035 (1976).

36. S. Ferrara and P. van Nieuwenhuizen, Phys. Rev. Lett. 37, 1669 (1976).

37. K. Johnson and E.C.G. Sudarshan, Ann. Phys. 13, 126 (1961).

38. G. Velo and D. Zwanziger, in Troubles in the External Field Problem for Invariant Wave Equations, A.S. Whightman ed., Gordon and Breach, N.Y., London, Paris 1971, Tract 4, p. 8.

39. S. Deser and P. van Nieuwenhuizen, Phys. Rev. Lett., 32, 245 (1974).

40. D.Z. Freeman and A. Das, Nucl. Phys. B120, 221 (1977).

41. S. Hawking, Cambridge University DAMTP preprint.

42. W. Nahm, Nucl. Phys. B135, 149 (1978).

43. L. Brink, J.H. Schwarz and J. Scherk, Nucl. Phys. B121, 77 (1977).

44. E. Cremmer and B. Julia, Phys. Lett. 80B, 48 (1978).

45. E. Cremmer and B. Julia, Nucl. Phys. B159, 141 (1979).

46. E. Cremmer in Proc. Europhysics Conf. on "Unification of the Fundamental Interactions" (Erice, 1980).

47. B. DeWit and D.Z. Freeman, Nucl. Phys. B130, 105 (1977).

48. J. Scherk and J.H. Schwarz, Phys. Lett. 82B, 60 (1979).
E. Cremmer, J. Scherk and J.H. Schwarz, Phys. Lett. 84B, 83 (1979).

49. E. Cremmer, S. Ferrara, K.S. Stelle and P.C. West, report LPTENS 80/13 (1980).

50. M.J. Duff, in Proc. Europhysics Conf. on "Unification of the Fundamental Interactions" (Erice 1980), to be pub.

51. Y. Ne'eman, Phys. Lett. B82, 69 (1979).

52. H. Harari, Phys. Lett. 86B, 83 (1979).
M.A. Shupe, Phys. Lett. 86B, 86 (1979).

53. J. Ellis, M. K. Gaillard, L. Maiani and B. Zumino, CERN report TH-2841 (1980).

54. J. Ellis, M.K. Gaillard and B. Zumino, CERN report TH 2842 (1980).

GROUP MANIFOLD APPROACH IN UNIFIED GRAVITY THEORIES

Tullio Regge

Istituto di Fisica Teorica dell'Università di Torino - Italia

The last decade has seen a revival of interest in unified models of elementary particles. I do not think that I am saying anithing new in emphasizing again the leading role of the concept of Yang-Mills fields in all these attemps. But perhaps it is interesting to see how Yang-Mills fields relate to modern differential geometry. By now everybody knows that they first appeared as "Cartan connection" in a generalization of Levi - - Civita parallel transport of lasting importance in modern geometry. In Cartan's mind however and in most contemporary applications of the idea we are dealing with pure geometry, there is no action nor any equation of motion in all the relevant applications or almost all of them. A notable exception is of course instanton theory where duality takes up to role of the equation of motion.

I would like to discuss here another possible role of differential geometry and perhaps of algebraic topology in unified theories. As I have hinted at Cartan's connection is geometry but only half of physics, in order to have a physical theory we need geometry plus an action principle. The struggle for general relativity was over once Hilbert and Einstein got their action. It is in general agreed that if you have a Yang Mills potential the best and only way to construct an action is to choose an action density which is an invariant under the relevant gauge group. This is not always a convenient or even a cor rect point of view. In dealing with conventional gravity in the vierbein formalism the object of the theory are 1-forms:

(1) $$V^a = V^a_\mu(x)\, dx^\mu, \quad \omega^{ab} = \omega^{ab}_\mu\, dx^\mu$$

From these forms we construct curvatures:

(2) $$R^a = \mathcal{D}V^a = dV^a - \omega^{ab}\wedge V_b, \quad R^{ab} = d\omega^{ab} - \omega^{at}\wedge\omega_t{}^b$$

Before proceeding further let me offer the following comments. It is possible to consider the whole potential multiplet as the potential μ^A for a local Poincare group and the corresponding curvatures as the field strengths, as in the conventional interpretation:

$$R^A = d\mu^A + \frac{1}{2} C^A{}_{\cdot BC}\, \mu^B \wedge \mu^C$$

However in gravity the action density is gives by:

$$\mathcal{L} = \text{const.}\ R^{ab}\wedge V^c \wedge V^d\, \epsilon_{abcd}$$

and is not invariant under the whole Poincare local gauge group. Alternatively one can consider as potentials only the spin connection and the Lorentz group as the only true gauge group. The action is now gauge invariant under a smaller group but is contains fields V^a which are no longer potentials. Here I am proposing to call a theory with only potentials as fields "pure" and "impure" otherwise. This nomenclature hints at an aesthetical criterion for a unified theory, pure theories are more economical and better than impure ones, they are "vernunftig" in the jargon of Einstein. So gravity is either pure but not invariant or impure but invariant, depending on the point of view. A similar comment holds for supergravity and other theories in higher dimensions. The structure of these theories is very interesting from the algebraic point of view. Since they are written in terms of forms they are de facto invariant under general coordinates transformations. But by now almost everyone agrees that there is no relevant physical meaning attached to coordinate transformations, they should be regarded merely as a relabeling of the points of space time or superspace. In this

context conventional coordinate transformations in general relativity are sometime confused with the missing Poincaré gauge traslational invariance, thus creating some confusion.
The problem is now that of defining in some unique way an action without recourse to invariance or using it only in part. How is the Einstein-Hilbert-Cartan action unique? What makes supergravity so special? What is the simplest way to characterize these theories? I offer here some proposal which should fit with the remarks of S. Sternberg at this conference.
To begin with I describe pure theories only. Through dimensional reduction we should get anyway from them a huge class of apparently impure theories in lower dimension. Also a pure theory may look impure if we choose to look at a subgroup H of the whole gauge group G as the "true" gauge group. With this convention gravity changes from pure to impure. It is not clear to me that all interesting impure theories can be obtained this way, this remains an outstanding algebraic and physical problem.
Furthermore the theory is written directly on the group manifold G.
We should understand that the gravity action density is integrated over a 4-manifold M imbedded in the Poincaré group or at least a manifold which is locally a diffemorph of the Poincaré group.
I neglect at the moment the otherwise extremely interesting interplay of global and local properties of these actions.
Finally we carry out variations in all relevant fields in a first order formalism on all possible choices of the imbedding of M. The equations read (for gravity):

$$R^a = 0 \qquad\qquad V^a R^{bc} \epsilon_{abcd} = 0 \quad \forall d \tag{3}$$

It turns out that, always for gravity, the role of the Lorentz group is here trivially factorizable and that only Minkowsky space remains as the true arena of physics. So the physical content of the group manifold approach is here the same as that of conventional space time formulation. A similar comment applies to supergravity and to other known pure theories, notably supergravity in 5 dimensions. I do not claim here that a conventional theory acquires a better look or a new physical content by writing it on the group manifold. Here we are doing something else, we are trying to classify "all" theories. Once we carry out the variation on G we discover that the number of field equations greatly exceeds the number of available components of the fields. Nevertheless two factors enter which reduce the number of actually independent equations. We have the Bianchi identities and the action has peculiar properties. Indeed if we consider supergravity:

$$(4)\quad \mathcal{L} = R^{ab}\wedge V^{c}\wedge V^{d}\,\epsilon_{abcd} + a\bar{\Psi}\wedge\gamma_5\gamma_a\rho\wedge V^{a} + b R^{ab}\wedge\bar{\Psi}\gamma_5\Sigma^{ab}\Psi$$

with an arbitrary constant a,b we discover the following facts:

1) If we choose a to be different from the value 4 it has in the standard theory we discover that the flat solution, superspace, is no longer solution of the equation of motion.
2) If the choose b $\neq$ 0 we discover that if now a = 4 the flat solution is the only solution of the theory, this becomes then trivial. A theory with only flat space as a solution (that is zero curvature) will be called "rigid" in the sequel. This conforms to the current mathematical usage where curved solutions are called "soft manifolds". Gravity in 3 dimensions is indeed rigid, so rigidity is not a new phenomenon. In general one starts from a generic gauge group G (containing the Poincaré or the De Sitter group) and proposes an action of the following sort:

$$(5)\quad A = \int \left(\Lambda + R^{A}\wedge\nu_{A} + R^{A}\wedge R^{B}\wedge\nu_{AB} + \ldots\text{etc.}\right)$$

Here the ν_A, ν_{AB} are polynomials in the connection and the R^A are the curvatures. The above action is the most generic one can form by using the exterior algebra on forms but without introducing the Hodge duality. As such it does not mention the dimension of the group, the degree of the form is of course the dimension of space time. Upon variation we obtain equations of the form:

$$(6)\quad 0 = \underline{A}\rfloor\Lambda + D\nu_{A} + (-1)^{AB} R^{B}\wedge\underline{A}\rfloor\nu_{B} + 2R^{B}\wedge D\nu_{AB} + (-1)^{A(B+C)} R^{B}\wedge R^{C} \times \underline{A}\rfloor\nu_{BC}$$

which are in general incompatible with the condition that flat space is a solution and not the only one. Imposing these requirements restricts greatly the action density and correspond to set a=4 and b=0 in supergravity. In dimension 5 the corresponding action is practically unique, in other dimensions and for

generic choices of the degree of the action there is no acceptable physical theory.
So although we have no invariance or a limited one the theory turns out to be practically unique.
How does it happen that the symmetry of the group G is broken? Clearly if the search for a lagrangian of the kind (5) has any success the theory must posses infinitely many equivalent lagrangians all differing by a trivial gauge transformation, of the kind which, indeed, does not leave the action invariant.
So the set of all possibile actions is invariant but a single choice is not, a structure which resembles spontaneous symmetry breaking. If a comment can be offered I could say that here breaking of the symmetry is even more spontaneous than the current one, we are facing an algebraic procedure which chooses a particular polynomial or rather a class of equivalent polynomials. The procedure is intrinsically invariant and has some similarity with the search for a Chevalley cohomology on the group manifold, in fact in some cases it even coincides with it. It is quite remarcable that invariance gets replaced by a purely algebraic condition and with equally restrictive effects.
In accepting this fact one must understand that the original class of actions (5) is quite restrictive, Maxwell action does not fit into it, it requires the dual of a form. This may seem a severe drawback. I do not thing so. Maxwell theory is retrieved in many pure theories through dimensional reduction as in the celebrated Klein-Kaluza example. It may be that this is the natural way to get it. Many people find it very disturbing that non abelian gauge theories cannot be naturally obtained this way in general they appear with very large cosmological constants and they refuse to conform to the requirement that flat space should be a solution. I feel confident that this obstacle and others will find a natural solution. Meanwhile the talk of Sterberg has been very timely, most probably the classification of action od the kind (5) should be carried out by using the concept of cathegory and the techniques of functors.
The condition that flat space should not be the only solution is very restrictive and is usually referred to as "rheonomy". If we start from an action which satisfies the usual criteria of invariance and the "vacuum" solution we find in general a rigid theory. On the other hand if we adjust the parameters in such a way as not to have a trivial theory we find that the condition for rheonomy is extremely restrictive and essentially

replaces invariance as a building device for the theory. There is no known general structure theorem which sets conditions for the existence of rheonomic theories in higher dimensions. The search for these theories is essentially equivalent to the search for supersymmetry along conventional methods.
It ha to be pointed out that variational equations derived from the group manifold approach fall in different cathegories. We have roughly the following classification.
1) Constraints equations. For instance the torsion equation for gravity:

$$R^a = 0$$

is usually interpreted as a constraints defining the spin connection in terms of the first order derivatives of the tetrad.
2) Equations of motion. Among these the Einstein field equations and the Rarita Schwinger equation for the gravitino in supergravity.
3) Rheonomic equations which implement the transformation rules for the relevant fields under supersymmetry. These equations are used to express the normal components of the forms, with respect to the world hypersurface in terms of the tangential ones. Once this relation is known it is possible to interpret the so called AGCT transformations:

$$\delta\mu^A = \nabla\varepsilon^A + C^A_{\cdot BC}\,\mu^B \wedge \varepsilon^C$$

as transformations of the fields.
Auxiliary fields appear in these theories in the following context. They are particular components of the curvature forms in the normal direction of the hypersurface M. It is possible to impose on these forms suitable conditions, compatible with the Bianchi identities, in such a way as to guarantee a unique transformation law under supersymmetries without implementing any of the equations of motion discussed in (2). One obtains in this way a structure which is intermediate in character between having the full equations on shell, that is all (1.2.3) referred to in the above, and the total absence of variational equation and no conditions imposed on the connection. As one can see in supergravity these structures are not unique, one can talk about a restricted or enlarged set of auxiliary fields. The chief use of auxiliary fields lie in the quantization of the theory. I suspect however that thay play a much deeper algebraic role in the whole set of theories described here. They

remain at present an unsolved riddle in that there is no established method for their definition and construction and they have to be worked out by hand in all existing theories.
Finally I would like to discuss many of the open problems of the group manifold approach. It still lacks a canonical formalism similar to the one developed for conventional gravity and supergravity, also it would be desirable to establish in the most general and elegant way a procedure for dimensional reduction.
One should obtain in this way the most powerful setting for the original ideas of Klein and Kaluza and derive also a large class of impure but interesting theories. Finally the problem of quantization is at the moment completely open and it has been examined only in the context of conventional theories. This problem, as we have stated, is directly related to the definition and use of auxiliary fields. In closing this discussion I would like to remark that the search for unified theories has barely begun and is far from being closed or anywhere near exhaustion. This should be an added incentive for further progress.

Acknowledgements - I would like to thank Dr. P. Fré and Dr. R. D'Auria for many illuminating conversations and contributions to the point of view expressed in this talk.

References - The interested reader should consult for details: "Group manifold approach to gravity and supergravity theories" by R. D'Auria, P. Fré and T. Regge, IC/81/54 IAEA preprint, Trieste.

EINSTEIN AND THE GEOMETRIZATION OF PHYSICS

Andrzej Trautman

Instytut Fizyki Teoretycznej
Uniwersytet Warszawski
Hoza 69, Warszawa
Poland

ABSTRACT: Einstein's programme of unification and geometrization of physics is briefly reviewed in the context of the current research on gauge theories of the Yang-Mills type.

1. The idea of connecting the motion of matter to the geometric properties of space -such as curvature - may be traced back to Lobacevskii [17], Riemann [20], Clifford [2] and Mach [18]. Einstein was the first to formulate a definite, precise, verifiable and verified theory which directly related the distribution and motion of matter to the geometry of spacetime. Moreover, his theory of general relativity reduces gravitation to geometry in the sense that all information about gravitational interactions is contained in the data on metrical relationships in spacetime. The simplicity and beauty of that theory encouraged mathematicians and physicists to look for its extension to electromagnetism, which, in the 1920s, was the only known, other than gravitation, domain of fundamental physical phenomena.

2. The search for a geometric and unified theory of physical fields occupied a major place in the scientific activity of Albert Einstein. Shortly after the general theory of relativity had been formulated (1915), Herman Weyl put forward a geometric model of gravitational and electromagnetic forces [26]. Einstein criticized [E1] the extension of Riemannian geometry proposed by Weyl, but, at the same time, he became fascinated with the idea of building a unified theory, providing a description of electromagnetic and gravitational fields in terms of differential geometry [E2].

3. Einstein's programme of constructing a unified theory has been often, and sometimes sharply, criticized by physicists. Einstein was accused of having ignored the growth of physics after 1935, when it became clear that there exist in nature fundamental interactions other than gravitational and electromagnetic forces. Moreover, all Einstein's attempts were in the framework of classical physics and he hoped to explain the discrete nature of matter within that scheme.

4. The original "Einstein Programme" of unifying physics can be

summarized as follows:

- construct a theory of a classical geometric field without sources;
- unify the basic variables of gravitation and electromagnetism into one geometric object;
- obtain unified field equations from a variational principle;
- describe charged particles as sourceless solutions and derive their motion from the field equations.

Einstein hoped that fundamental properties of elementary particles and their apparent quantum behaviour could also be somehow explained on the basis of a geometric, classical and unified theory.

5. The relativistic theory of gravitation is based on two fundamental geometric objects: a metric tensor g and a linear connection Γ. The metric is needed to measure distances, time intervals, relative velocities and angles. Based on the notion of parallel transport, due to Levi-Civita, the connection is a somewhat subtler concept needed to compare directions, forces and fields at points separated in space and time.

6. Essentially all attempts at unification rely on geometrics based on g and Γ. In order to obtain new degrees of freedom needed to describe electromagnetism, one either relaxes the restrictions imposed on g and Γ in Riemannian geometry, or increases the number of dimensions of the underlying manifold, or both.

7. The first 'unified theory' was due to Weyl [26]: he observed that Einstein's theory of gravitation is based on a 'relativity of directions' and proposed to extend it to account for a 'relativity of magnitude' by allowing conformal transformations of the metric,

$$g'_{\mu\nu} = e^{f} g_{\mu\nu} \tag{1}$$

If

$$\nabla_\rho g_{\mu\nu} = A_\rho g_{\mu\nu}$$

then Eq. (1) is incompatible with the gauge transformation

$$A'_\mu = A_\mu + \partial f/\partial x^\mu$$

Einstein criticized Weyl's theory by pointing out that, according to (1), $ds^2 = g_{\mu\nu}dx^\mu dx^\nu$ is not well-defined and the rate of atomic clocks may depend on their history. Weyl's theory played a major role in the history of physics [30]: in 1929 Weyl interpreted gauge transformations as acting on wavefunctions of charged particles, rather than on g . This idea led later to non-Abelian gauge theories [29] and to the interpretation of electromagnetic and Yang-Mills potentials as connections on principal fibre bundles (cf. § 12).

8. A class of unified theories, originated by Kaluza [8] and Klein [12], has been developed by Einstein and his coworkers [E3, E10, E26-29] and other authors [6, 15, 22]. These theories assume a 5-dimensional Riemannian geometry of a special type. It has been recognized that the Riemannian metric on the Kaluza-Klein space is equivalent to the metric which may be introduced, in a natural manner, on the total space P of a U(1)-bundle over spacetime M, provided that M has a Riemannian metric and P carries a U(1)-connection, i.e. an electromagnetic potential. This observation has been generalized to non-Abelian gauge configurations [1, 4, 9, 13, 24].

9. Eddington [4] proposed to consider Γ as a basic quantity and to derive from it both the metric tensor and the electromagnetic field by splitting $R_{\mu\nu}$ into its symmetric and skewsymmetric parts. Einstein developed this idea by postulating that the Lagrange density should be proportional to $|\det R_{\mu\nu}|^{\frac{1}{2}}$. Unfortunately, this led to equations incompatible with experiments and Einstein abandoned this approach [E4-8]. Very recently, Kijowski and Tulczyjew were able to improve the Eddington-Einstein theory and to derive equations equivalent to the Einstein-Klein-Gordon and the Einstein-Maxwell systems without introducing the metric tensor as a fundamental variable [10,11].

10. The theories based on teleparallelism assume a vanishing curvature of the connection Γ, compatible with the metric tensor g. In other words, all geometry - and physics, so long as only gravitation and electromagnetism are considered - is presumed to be contained in g and the torsion tensor Q. These theories, developed by Einstein in collaboration with Mayer [E11-23, E25] also turned out to be non-viable, but they now play a role in the study of 'gauge theories of gravity', developed by Hehl, Ne'eman, Von der Heyde and their coworkers [5].

11. In 1925 Einstein briefly considered [E9] a theory based on Γ and a nonsymmetric density $g^{\mu\nu}$; he identified $g^{[\mu\nu]}$ with the electromagnetic field. He returned to this idea in 1945 [E22-E42] and spent the last ten years of his life working on an 'asymmetric theory' based on $g_{\mu\nu}$ and $\Gamma^{\rho}_{\mu\nu}$, both of which need not be symmetric in the pair (μ,ν). Schrödinger [21] and many other authors [7, 23] contributed to this field of research, but not successful, unified theory emerged out of their work.

12. The development of physics in recent years throws a new light on the idea of geometrizing physics. It seems that all fundamental interactions exhibit similarities that are most easily perceived and described in the language of differential geometry. There is a prospect for achieving a unification of weak and electromagnetic interactions (the Weinberg-Salam theory). According to Chen Ning Yang [31],

> "Einstein's insistence on the importance of unification, was a deep insight, which he courageously defended, against all

spoken and unspoken criticism... . It turns out that the structure that Einstein was seeking was the gauge field".

Indeed, it appears that the idea of a connection has a significance which goes well beyond the theory of gravitation. The potential of an electromagnetic field plays a role in comparing the phases of wave functions of charged particles at different points. This leads to the important idea of a gauge field. Generalized by Yang and Mills to more complicated "phase factors" which can change by means of transformations belonging to a non-Abelian group, gauge fields have become the most promising candidate to provide a geometric description of physics. From the point of view of mathematics they are connections on principal fibre bundles.

13. Gauge theories attract much attention because they are renormalizable. Moreover, the non-Abelian theories allow a spontaneous breaking of symmetries that leads to massive vector particles, needed to describe the short-range character of nuclear forces.

14. If the current views are confirmed, then the four fundamental interactions that (probably) underlie all physical phenomena will allow a description by means of gauge fields - or, equivalently, connections - associated with Lie groups. In the case of gravitation, the Lorentz group plays the fundamental role, the Weinberg-Salam theory is based on $U(1) \times SU(2)$, whereas the current model of strong interactions (quantum chromodynamics) relies on $SU(3)$.

15. If the future development of physics confirms the hopes physicists now associate with gauge fields, then connections on principal bundles will provide the key to a geometrization of physics close in spirit, if not in detail, to Einstein's dream.

ACKNOWLEDGEMENT

The lecture was prepared during my stay, in May 1980, at the State University of New York at Stony Brook, N.Y.

I thank Professors C. Denson Hill, Irwin Kra and Chen Ning Yang for their hospitality and discussions. I have also been influenced by conversations with Engelbert Schuecking held in Washington Square, New York, N.Y., and with Yuval Ne'eman enjoyed in Erice, Sicily.

EINSTEIN'S PAPERS ON UNIFIED FIELD THEORIES

This list is as complete as the author of the article has been able to make it. Many of Einstein's papers have been published in the Sitzungsberichte Preuss. Akademie der Wissenschaften (Berlin). The name of this journal is abbreviated here to SB. At the end of most listings there is a code letter relating the paper to one of five main tracks in the fields of unified theories. The meaning of the code letters is as follows:

A = asymmetric theories, i.e. theories with an asymmetric metric tensor and a linear connection with torsion,
B = pure affine theories, initiated by Eddington and based on a symmetric linear connection,
K = theories of the Kaluza-Klein type,
T = theories with teleparallelism,
W = the Weyl theory.

References in the text to Einstein's papers listed below are given in the style [En].

1. Eine naheliegende Ergänzung des Fundamentes der allgemeinen Relativitätstheorie, SB (1921), 261-264 (W).
2. Grundgedanken und Probleme der Relativitätstheorie, in Nobelstiftlesen, Les Prix Nobel en 1921-22, Imprimerie Royale, Stockholm 1923 (general remarks on the programme of unification).
3. (with J. Grommer) Beweis der nichtexistenz eines überall regulären zentrisch symmetrischem Feldes nach der Feldtheorie von Th. Kaluza, Jerusalem University, Scripta 1 (1923), No. 7 (K).
4. Zur allgemeinen Relativitätstheorie, SB (1923), 32-38 (B).
5. Bemerkung zu meiner Arbeit «Zur allgemeinen Relativitätstheorie», SB (1923), 76-77 (B).
6. Zur affinen Feldtheorie, SB (1923), 137-140 (B).
7. The theory of the affine field, Nature 112 (1923), 448-449 (B).
8. (Anhang:) Eddingtons Theorie und Hamiltonsches Prinzip, in the book by A.S. Eddington «Relativitätstheorie in mathematischer Behandlung», Springer, Berlin, 1925 (B).
9. Einheitliche Feldtheorie von Gravitation und Elektrizität, SB (1925), 414-419 (A).
10. Zu Kaluzas Theorie der Zusammenhängs von Gravitation und Elektrozität, SB (1927), 23-25 and 26-30 (K).
11. Riemann-Geometrie mit Aufrechterhaltung der Begriffes des Fernparallelismus, SB (1928), 217-221 (T).
12. Neue Möglichkeit für eine einheitliche Feldtheorie von Gravitation und Elektrizität, SB (1928), 224-227 (T).
13. Über den gegenwartigen Stand der Feldtheorie, «Festschrift Prof. Dr. A. Stodola zum 70. Geburstag», Füssli Verlag, Zürich, 1929 (T).
14. Zur einheitlichen Feldtheorie, SB (1929), 2-7 (T).
15. The new field theory, Observatory 52 (1929), 82-87 and 114-118 (T).
16. (with Th. de Donder) Sur la théorie synthétique des champs, Revue gén. de l'électricité 25 (1929), 35-39 (T).
17. Einheitliche Feldtheorie und Hamiltonsches Prinzip, SB (1929), 156-159 (T).
18. Theorie unitaire de champ physique, Ann. Inst. H. Poincaré 1 (1930), 1-24 (T).
19. Auf fie Riemann-Metrik und den Fern-Parallelismus gegründete einheitliche Feldtheorie, Math Ann. 102 (1930), 685-697 (T).

20. Die Kompatibilität der Feldgleichungen in der einheitlichen Feldtheorie, SB (1930), 18-23 (T).
21. (with W. Mayer) Zwei strenge statische Lösungen der Feldgleichungen der einheitlichen Feldtheorie, SB (1930), 110-120 (T).
22. Zur Theorie der Räume mit Riemann-Metrik und Fernparallelismus SB (1930), 401-402 (T).
23. Uber den gegenwärtigen Stand der allgemeinen Relativitätstheorie, Yale Univ. Library Gazette 6 (1930), 3-6 (T).
24. Gravitational and electrical fields, Science 74 (1930), 438-439 (K).
25. (with W. Mayer) Systematische Untersuchunger über kompatible Feldgleichungen, welche in einem Riemannschen Raume mit Fernparallelismus gesetzt werden können, SB (1931), 252-265 (T).
26. (with W. Mayer) Einheitliche Theorie von Gravitation und Elektrozität, SB (1931), 541-557 and (1932) 130-137 (K).
27. Der gegenwärtiger Stand der Relativitätstheorie, Die Quelle (Pëdagogisher Führer) 82 (1932), 440-442 (K).
28. (with P.G. Bergmann) Generalization of Kaluza's theory of electricity, Ann. of Math. 39 (1938), 683-701 (K).
29. (with V. Bargmann and P.G. Bergmann) On five-dimensional representation of gravitation and electricity, in: Th. von Karman Anniversary Volume, pp. 222-225, Caltech, Pasadena, 1941 (K).
30. (with V. Bargmann) Bivector fields I, Ann. Math. 45 (1944), 1-14.
31. Bivector fields II, Ann. Math. 45 (1944), 15-23.
32. Generalization of the relativistic theory of gravitation, Ann. Math. 46 (1945), 578-584 (A; g and Γ complex).
33. (with E. Straus) Generalization of the relativistic theory of gravitation II, Ann. Math. 47 (1946), 731-741 (A).
34. Generalized theory of gravitation, Rev. Mod. Phys. 20 (1948), 35-39 (A).
35. On the generalized theory of gravitation, Sci. Amer. 182 (1950), 13-17 (A).
36. The Bianchi identities in the generalized theory of gravitation, Cand. J. Math. 2 (1950), 120-128 (A).
37. Generalization of the theory of gravitation, Appendix II to «The Meaning of Relativity»,fourth edition, Princeton, 1953 (A).
38. A comment on a Criticism of unified field theory, Phys. Rev. 89 (1953), 321 (A).
39. (with B. Kaufmann) Sur l'état actuel de la théorie générale de la gravitation, in «Louis de Broglie, Physicien et Penseur» Albin Michel, Paris 1952; there is in this book an Appendix by A. Einstein, Extension du groupe relativiste, devoted to λ-transformations (A).
40. (with B. Kaufmann) Algebraic properties of the field in the relativistic theory of the asymmetric field, Ann. Math. 59 (1954), 230-244 (A).
41. (with B. Kaufmann) A new form of the general relativistic field equations, Ann. Math. 62 (1955), 128-138 (A).
42. **Relativistic theory of the non-symmetric field, Appendix II to «The Meaning of Relativity», fifth edition, Princeton, 1955 (A).**

OTHER REFERENCES

References in the text to the papers and books listed below are indicated by appropriate numbers in square brackets.

1. Cho, Y.M., J. Math. Phys. 16 (1975), 2029.
2. Clifford, W.K., On the space-theory of matter, published for the first time in 1876, reprinted in «Mathematical Papers», MacMillan, New York and London, 1968.
3. De Donder, Th., La gravifique de Weyl-Eddington-Einstein. Gauthier-Villars, Paris, 1924.
4. DeWitt, B.S., in «Relativité, groupes et topologie» p. 725, ed. by C. DeWitt and B.S. DeWitt, Gordon and Breach, New York, 1964.
5. Eddington, A.S., Proc. Roy. Soc. (London) A99 (1921), 104.
6. Hehl, F.W. Four lectures on Poincaré gauge field theory, in: Cosmology and Gravitation, edited by P.G. Bergmann and V. De Sabbata, Lectures at Erice (May 1979), Plenum Press, New York, 1980.
7. Hlavaty, V., Geometry of Einstein's unified field theory, Noordhoff, Groningen, 1956.
8. Kaluza, Th., SB (1921), 966.
9. Kerner, R., Ann. Inst. H. Poincaré 9 (1968), 143.
10. Kijowski, J., GRG Journal 9 (1978), 857.
11. Kijowski, J., and W.M. Tulczyjew, A symplectic framework for field theories, Lecture Notes in Physics No. 107, Springer-Verlag, Berlin, 1979.
12. Klein, O., Z. Phys. 37 (1926), 895.
13. Kopczyński, W., Acta Phys. Polon. B 10 (1979), 365 and in «Differential Geometric Methods in Mathematical Physics» p. 462, ed. by P. Garcia et al., Lecture Notes in Mathematics No. 836, Springer-Verlag, Berlin, 1980.
14. Levi-Civita, T., A simplified presentation of Einstein's unified equations, Blackie and Son, London, 1929.
15. Lichnerowicz, A., Théorie relativistes de la gravitation et de l'électromagnétisme, Masson, Paris, 1955.
16. Lichnerowicz, A., Differential geometry and physical theories, in: «Perspectives in Geometry and Relativity» (Essays in honor of Vaclav Hlavaty) ed. by B. Hoffmann, Indiana Univ. Press, Bloomington, 1966.
17. Lobačevskii, N.I., Exposition succincte des principes de la géométrie. Lecture given on 12 February 1826, reprinted in «Osnovaniya geometrii» (Foundations of geometry), GITTL, Moscow, 1956.
18. Mach, E. Die Mechanik in ihrer Entwicklung historisch-kritisch dargestellt, Brockhaus, Leipzig, 1904.
19. Penrose, R., The geometry of the universe, in «Mathematics Today», ed. L.A. Steen, Springer-Verlag, New York, 1979.
20. Riemann, B., On the hypotheses that lie at the foundations of geometry. Habilitationsvorlesung of 10 June 1854, in «Gesammelte Mathematische Werke», 2nd ed., ed. by H. Weber, Dover, New York, 1953.

21. Schrödinger, E., Proc. Roy. Irish Acad. 51A (1946), 41.
22. Thirring, W., Remarks on five-dimensional relativity, in «The Physicist's Conception of Nature», ed. by J. Mehra, D. Reidel, Dordrecht, 1973.
23. Tonnelat, M.A., Les théories unitaires de l'électromagnetisme et de la gravitation, Gauthier-Villars, Paris, 1965.
24. Trautman, A., Lectures at King's College, London, 1967; published in Rep. Math. Phys. (Toruń) 1 (1970), 29.
25. Trautman, A., Fibre bundles, gauge fields and gravitation, in: «General Relativity and Gravitation», Vol. I pp. 287-308, A. Held et al., eds., Plenum Press, New York, 1980.
26. Weyl, H., SB (1918), 464; Ann. d. Phys. (Leipzig) 59 (1919), 101.
27. Weyl, H., Z. Phys. 56 (1929), 330.
28. Wheeler, J.A., Einsteins Vision, Springer-Verlag, Berlin, 1968.
29. Yang, C.N. and R.L. Mills, Phys. Rev. 96 (1954), 191.
30. Yang, C.N., Ann. N.Y., Acad. Sci. 294 (1977), 86.
31. Yang, C.N., Lecture at 2nd M. Crossmann Meeting, Trieste, July 1979 and Physics Today, June 1980, pp. 42-49.
32. Zaycoff, R., Z. Phys. 65 (1931), 428 and 67 (1931), 135.

CLASSICAL SCATTERING THEORY

W. Thirring

Institut für Theoretische Physik

Universität Wien, Austria

1. Introduction

It was first recognized by Hunziker [1] that the notions of scattering theory play an important role in classical mechanics. It turned out [2] that it leads to non-trivial information for the global properties of the solutions of the classical trajectories. For instance it shows that in the three body problem there are large regions in phase space with $2n - 1 = 17$ constants of motion and all trajectories in this region are homotopic to straight lines. Furthermore Wigner's [3] time delay has a simple geometrical meaning [4] for the trajectories. Recently Bollé and Osborn [5] succeeded in deriving even a classical analogue to Levinson's theorem. In these lectures I shall, following [6], show in detail how the phase shift corresponds to the generator of the S-transformation. For this purpose we define in the next section canonical coordinates for a one-, two- and three-dimensional configuration space such that this statement assumes a simple form. This sheds some light on how trajectories with large time delays or loopings generate resonances of the quasiclassical phase shift. In the following section we give an alternative proof of the classical form of Levinson's theorem and illustrate its subtle feature by some examples. Finally we give a simple derivation of how a Dollard's [7] change in the free motion leads to the Coulomb phase shift as generator of the classical S-transformation for a 1/r-potential.

We shall employ the following

Notations:

$$\Theta(x) \equiv \begin{matrix} 1 & \text{for} & x > 0 \\ 0 & \text{for} & x < 0 \end{matrix} \qquad \text{(step function)}.$$

$f \circ g(x) = f(g(x))$ (composition of maps).

$\sup\limits_{x} f(x)$ = least upper bound of f.

$\vec{a} \wedge \vec{b}$ = vector product.

2. The S-transformation

Scattering theory investigates the asymptotic behaviour of the trajectories in phase space. Only some observables, typically functions of the momentum, will converge for $t \to \pm \infty$. This suffices to define the scattering angle

$$\theta = \sphericalangle (p_+, p_-) , \qquad p_\pm = \lim_{t \to \pm\infty} p(t) ,$$

but if the potentials decrease faster than 1/r for $r \to \infty$ additional information becomes available. Then time evolution $\Phi_t : (x(0), p(0)) \to$ $\to (x(t), p(t))$ though not tending to a limit for $t \to \pm \infty$ approaches the free time evolution Φ^0_t such that $\Phi_{-t} \circ \Phi^0_t$ tends to a limit in some regions $D_\pm$

$$\Omega_\pm = \lim_{t \to \pm\infty} \Phi_{-t} \circ \Phi^0_t .$$

This implies the convergence of $\vec{x}(t) - \vec{p}(t).t$ and allows the definition of the time delay compared with the free motion. Since Φ and Φ^0 are canonical transformations the "Möller-transformations" $\Omega_\pm$ will in general be a local canonical transformation mapping the domains $D_\pm$ into ranges $R_\pm$. Closed orbits will be excluded from $R_\pm$ but in reasonable [8] cases their union with R_+ or R_- will fill almost all of phase space. The scattering transformation

$$S = \Omega_+^{-1} \circ \Omega_- = \lim_{t \to \infty} \Phi^0_{-t/2} \circ \Phi_t \circ \Phi_{-t/2} \qquad (1)$$

transforms D_- into D_+. If Φ_t^o is the free time evolution having straight lines as trajectories, S has a simple geometrical meaning. For negative t $\Phi_{-t} \circ \Phi_t$ means that you follow the straight trajectory back for a time $-|t|$ and then continue with the actual time evolution for the same length of time. If the forces have a finite range and Φ and Φ^o coincide outside a certain region then $\Phi_{-t} \circ \Phi_t^o$ will become independent of t as soon as t leads you outside this region (Fig. 1). Then the limit is attained, Ω_- maps the straight line onto this trajectory of Φ_t which is asymptotically tangent to it. Similar arguments for Ω_+ show that S maps (Fig. 1) the straight lines tangent for $t \to -\infty$ onto the ones tangent for $t \to +\infty$. It follows from its definition that S commutes with the free time evolution: $S \circ \Phi_t^o = \Phi_t^o \circ S$. As a canonical transformation one should be able to exhibit its generator which actually is possible. We first study the special cases.

Examples

a) One-dimensional motion:

Let (x,p) be the canonical variables and consider the motions due to $H^o = p^2/2$, $H = p^2/2 + V(x)$ where V(x) has finite range or falls off sufficiently fast. The corresponding flows are

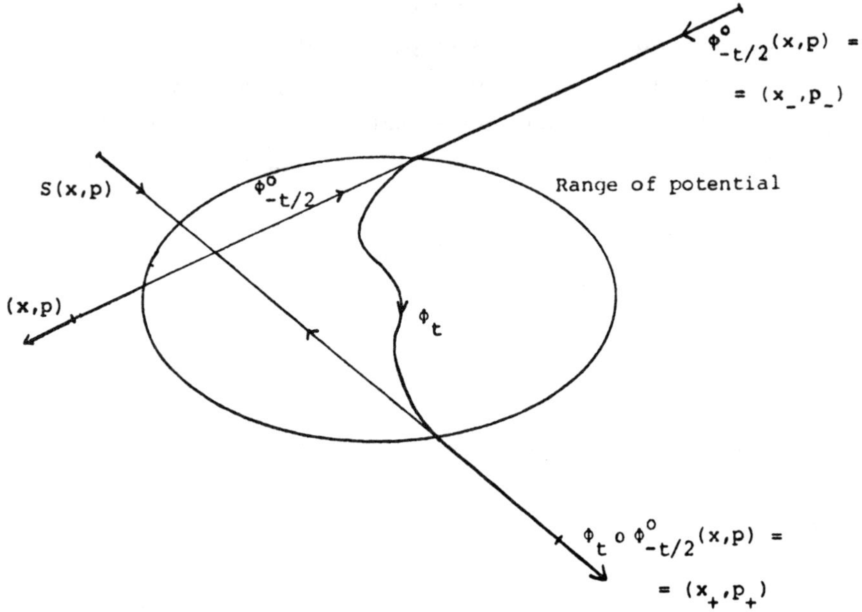

Fig. 1. Motion in configuration space

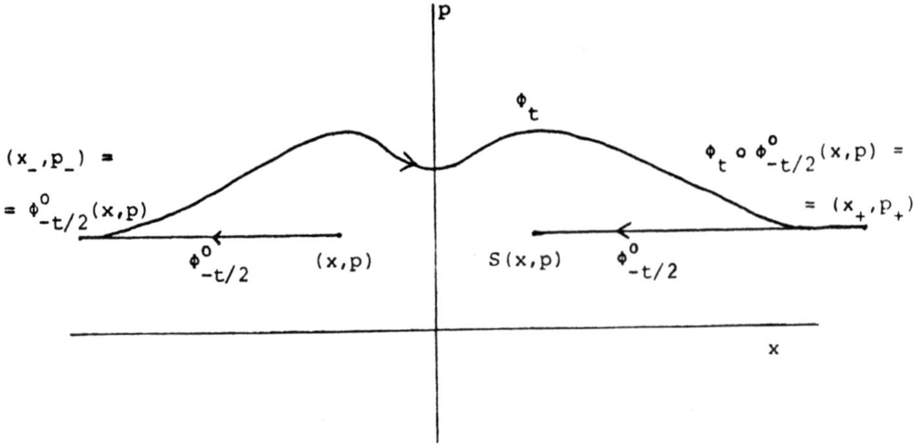

Fig. 2. One-dimensional motion in phase space

$$\Phi_t^o: (x,p) \to (x+pt,p)$$

$$\Phi_t(x,p) \to (x(t),\sqrt{p^2 + 2(V(x) - V(x(t)))} \quad . \tag{2}$$

If $p^2 > \sup_x 2V(x)$, so that there is no reflection, $x(t)$ is determined by

$$t = \int_x^{x(t)} \frac{d\alpha}{\sqrt{p^2 + 2(V(x) - V(\alpha))}} \quad .$$

In this case the S-transformation can be easily constructed: If we call $x_- = \Phi^o_{-t/2}(x) = x - pt/2$, $x_+ = \Phi_t(x_-)$ then $\Phi^o_{-t/2} \circ \Phi_t \circ \Phi^o_{-t/2}$ acts as

$$(x,p) \xrightarrow{\Phi^o_{-t/2}} (x - pt/2,p) \xrightarrow{\Phi_t} (x - pt/2 + \int_{x_-}^{x_+} d\alpha, \sqrt{p^2+2V(x_-)-2V(x_+)})$$

$$\xrightarrow{\Phi^o_{-t/2}} (x - \frac{t}{2}(p + \sqrt{p^2+2V(x_-)-2V(x_+)}) + \int_{x_-}^{x_+} d\alpha, \sqrt{p^2+2V(x_-)-2V(x_+)}) \tag{3}$$

where

$$t = \int_{x_-}^{x_+} \frac{d\alpha}{\sqrt{p^2 + 2(V(x_-) - V(\alpha))}} \quad .$$

For $t \to \infty$ we have $x_- \to -\infty$, $x_+ \to +\infty$ and $V(x_-)$, $V(x_+) \to 0$. Then

$$(x,p) \overset{S}{\to} (x - p \int_{-\infty}^{\infty} d\alpha(\frac{1}{\sqrt{p^2- 2V(\alpha)}} - \frac{1}{\sqrt{p^2}}),p) \equiv (x-p\tau,p) \tag{4}$$

i.e. S changes x by p times the time delay τ. The latter is the difference of the times the actual and the free time evolutions need for the trajectory from (x_-,p_-) to (x_+,p_+) in the limit $x_- \to -\infty$, $x_+ \to +\infty$ (Fig. 2). S is the canonical transformation $(x,p) \to (x - 2\partial\delta(p)/\partial p,p)$ where

$$\delta(p) = \frac{1}{2} \int_{-\infty}^{\infty} d\alpha(\sqrt{p^2 - 2V(\alpha)} - \sqrt{p^2}) \quad .$$

If there is an x_1 such that $p^2 < 2V(x_1)$ then a trajectory with $x < x_1$, $p > 0$ will be reflected to the left. In this case the action (3) of S is changed to ($V_- \equiv V(x_-)$, etc.)

$$(x,p) \xrightarrow{\Phi^o_{-t/2}} (x - pt/2,p) \xrightarrow{\Phi_t} (x_+,-\sqrt{p^2+2V_--2V_+}) \xrightarrow{\Phi^o_{-t/2}}$$

$$\to (x_+ + \frac{t}{2}\sqrt{p^2+2V_--2V_+},-\sqrt{p^2+2V_--2V_+}) \quad .$$

Here

$$t = \int_{x_-}^{x_o} \frac{d\alpha}{\sqrt{p^2+2V_--2V(\alpha)}} + \int_{x_+}^{x_o} \frac{d\alpha}{\sqrt{p^2+2V_--2V(\alpha)}}$$

where the reflection point x_o is the smallest x with $2V(x) = p^2 + 2V_-$. If $x_o < 0$ we may write

$$x_+ = - x_- - \int_{x_+}^{x_o} d\alpha - \int_{x_o}^{-x_o} d\alpha - \int_{-x_o}^{-x_-} d\alpha \quad ,$$

and thus for $t \to \infty$ we obtain

$$(x,p) \overset{S}{\to} (\lim_{t\to\infty} (- x + pt/2 - \int_{x_+}^{x_o} d\alpha - \int_{x_o}^{0} d\alpha - \int_{-x_o}^{-x_-} d\alpha + \frac{t}{2}\sqrt{p^2+2V_--2V_+}),\ -\sqrt{p^2+2V_--2V_t}) =$$

$$= (- x + 2p \int_{-\infty}^{x_o} d\alpha \{\frac{1}{\sqrt{p^2-2V(\alpha)}} - \frac{1}{\sqrt{p^2}}\} - \int_{x_o}^{-x_o} d\alpha,\ -p) \quad .$$

Thus the time delay is in comparison with a free motion going up to the origin. If V is twice differentiable it becomes infinite when p^2 approaches $2 \sup_x V(x) = 2V(x_o)$ because

$$\int \frac{d\alpha}{\sqrt{V''\cdot(x_o-\alpha)^2}} = \infty \quad .$$

Then there is an orbit which approaches x_o for $t \to \infty$ and for this value of p S does not exist. It separates the region in phase space where S is of the form (3) and the present case (Fig. 3) where

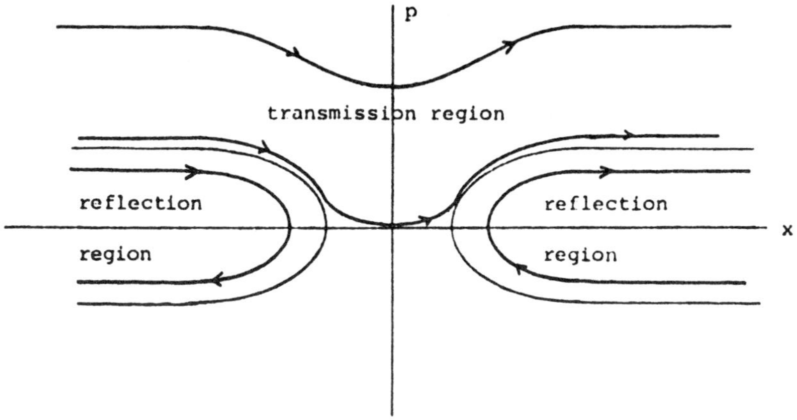

Fig. 3. Trajectories in phase space for a repulsive potential

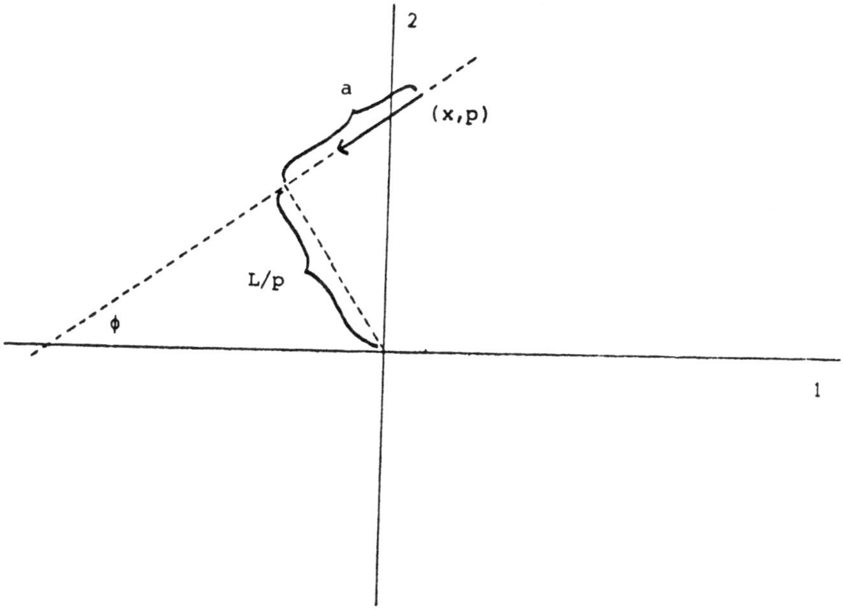

Fig. 4. The canonical coordinates in 2 dimensions

$$(x,p) \overset{S}{\rightarrow} (-x + 2\,\frac{\partial\delta(p)}{\partial p},\, -p)$$

$$\delta(p) = \int_{-\infty}^{x_0(p)} d\alpha(\sqrt{p^2-2V(x)} - \sqrt{p^2}) - p|x_0(p)| \quad .$$

For example the square well potential, $V(x) = V\,\Theta(R - |x|)$, gives

$$\delta(p) = R(\sqrt{p^2 + 2|V|} - \sqrt{p^2}) \quad \text{for} \quad V < 0$$

$$= R(\sqrt{p^2 - 2V} - \sqrt{p^2}) \quad \text{if} \quad p^2 > 2V$$

$$\qquad\qquad\qquad\qquad \text{for } V > 0 \quad .$$

$$= -\,pR \quad \text{if} \quad p^2 < 2V$$

b) Two-dimensional motion

Consider again $H^o = \frac{\vec{p}^2}{2}$, $H = \frac{\vec{p}^2}{2} + V(x)$. It is convenient to use the variables $p = |\vec{p}|$, $a = \frac{\vec{p}\cdot\vec{x}}{p}$, $L = |\vec{x} \wedge \vec{p}|$ and $\chi = \text{arc cos } p_x/p$, i.e. (Fig. 4)

$$p_x = p \cos\chi \ , \qquad x = a\cos\chi - \frac{L}{p}\sin\chi \ ,$$

$$p_y = p \sin\chi \ , \qquad y = a\sin\chi + \frac{L}{p}\cos\chi \quad .$$

Since L generates a rotation and $(\vec{p}\cdot\vec{x})$ the dilation $(x,p) \rightarrow (x\,e^{\beta}, p\,e^{-\beta})$ one sees readily that the new variables are canonical, i.e. $\{a,p\} = \{\chi,L\} = 1$, the other Poisson brackets vanishing. The free time evolution is simply $(a,\chi;p,L) \rightarrow (a + pt,\chi;p,L)$ but in general Φ_t will be complicated. S maps free trajectories into free trajectories and will be of the form

$$(a,\chi;p,L) \overset{S}{\rightarrow} (a - p\tau, \chi'; p, L') \quad ,$$

χ' and L' independent of a. If V is a central potential $V(|\vec{x}|)$ or more generally of the form $V(|\vec{x}|,L)$ so that L is constant, then Φ_t can be reduced to a one-dimensional problem and S constructed explicitly. The chain (3) of maps becomes in that notation

$$(a,\chi;p,L) \xrightarrow{\Phi^0_{-t/2}} (a - pt/2,\chi,p,L) \xrightarrow{\Phi_t} (a - pt/2 + \int_{a_-}^{a_+} da,\chi';\sqrt{p^2+2(V_- - V_+)},L) \rightarrow$$

$$\xrightarrow{\Phi^0_{-t/2}} (a - \frac{t}{2}(p + \sqrt{p^2+2(V_- - V_+)}) + \int_{a_-}^{a_+} d\alpha,\chi',\sqrt{p^2+2(V_- - V_+)},L) \quad . \qquad (5)$$

Again for $t \rightarrow \infty$ we have $|a_+|$, $|a_-| \rightarrow \infty$, V_+, $V_- \rightarrow 0$. The time t for the actual motion Φ_t is readily expressed as integral over $r \equiv |\vec{x}| = \sqrt{a^2+L^2/p^2}$ with $E = p^2/2 + V_-$

$$t = \int_{r_0}^{r_-} \frac{dr}{\sqrt{2E - L^2/r^2 - 2V(r)}} + \int_{r_0}^{r_+} \frac{dr}{\sqrt{2E - L^2/r^2 - 2V(r)}}$$

where $\sqrt{\ } = 0$ for $r = r_0$. Since $a = \sqrt{r^2 - L^2/p^2} = r + O(1/r)$ and $da =$ $= p\, dr/\sqrt{p^2 - L^2/r^2}$ we have for r_+, $r_- \rightarrow \infty$:

$$\int_{a_-}^{a_+} d\alpha = (\int_{L/p}^{r_-} + \int_{L/p}^{r_+}) \frac{p\, dr}{\sqrt{p^2 - L^2/r^2}} \quad .$$

Thus in the limit $t \rightarrow \infty$ we arrive at

$$(a,\chi;p,L) \overset{S}{\rightarrow} (a - 2\frac{\partial\delta(p,L)}{\partial p},\chi - 2\frac{\partial\delta(p,L)}{\partial L},p,L) \quad ,$$

$$\delta(p,L) = \lim_{R\rightarrow\infty} (\int_{r_0}^{R} dr \sqrt{p^2 - L^2/r^2 - 2V(r)} - \int_{L/p}^{R} dr \sqrt{p^2 - L^2/r^2}) \quad . \qquad (6)$$

Here we have used the well-known expression for the scattering angle $\chi-\chi'$.

Let us consider the two-dimensional S-transformation for typical potentials.

<u>$1/r^2$-potential</u>:

If $V(r) = c/r^2$ then S exists only for $c > - L^2/2$. In the attractive case trajectories with the impact parameter and therefore L too small, spiral into the origin. For the others the r-integral is easily calculated, for instance by complex integration. Evaluating the residue at the origin we find

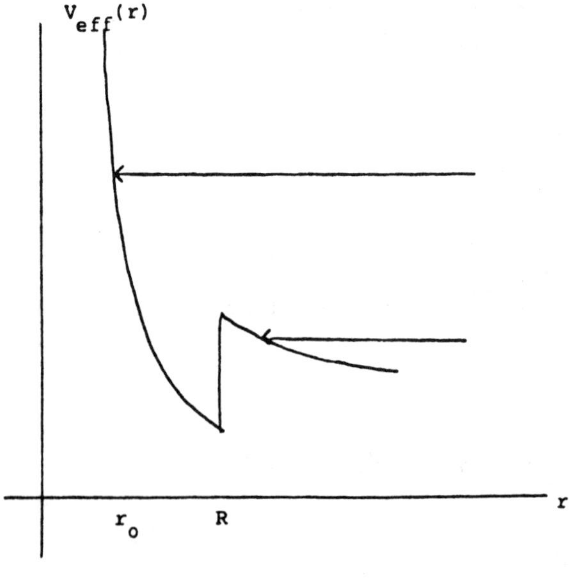

Fig. 5. The effective square well potential

$$\delta(p,L) = \frac{\pi}{2}\,[\sqrt{L^2+2c} - L] \quad .$$

Since δ is independent of p the time delay is zero. This is related to the dilation properties of the Hamiltonian (compare [2, 3.4.15.3]. The scattering angle $\pi(\frac{L}{\sqrt{L^2+2c}} - 1)$ tends to ∞ for $c < 0$ for those L where spiral orbits set in.

Square well potential:

If $V(r) = c\,\Theta(R-r)$ we find

$$\delta(p,L) = \{\sqrt{(p^2-2c)R^2-L^2} + L \arcsin \frac{L}{R\sqrt{p^2-2c}}\}\ \Theta(p^2R^2-L^2)\Theta((p^2-2c)R^2 - L^2) -$$

$$- \{\sqrt{p^2R^2-L^2} + L \arcsin \frac{L}{Rp}\}\ \Theta(p^2R^2 - L^2) \quad .$$

In the attractive case δ is discontinuous for $p^2 = 2c + L^2/R^2$ because there r_o jumps from $L/\sqrt{p^2-2c}$ to L/p (see Fig. 5).

Thus δ is given in the two regions $p^2 \lessgtr 2c + L^2/R^2$ by different expressions whereas on the separating hypersurface S does not exist. For a rounded off potential there are the trajectories which asymptotically reach the maximum of the potential but never get over it. For a twice differentiable potential the time delay would become ∞ whereas for the square well

$$\tau = \frac{2}{p}\frac{\partial\delta}{\partial p} = \frac{2}{p^2}\,\{\sqrt{p^2R^2 - L^2 - 2cR^2} - \sqrt{p^2R^2 - L^2}\}$$

remains finite. In any case this is the closest similarity to a quantum mechanical resonance since the special value $\delta/\hbar = \pi/2$ has classically no significance.

c) Three-dimensional motion

There are many canonical coordinate systems such that the free motion just shifts one coordinate. We shall choose one where $|\vec{p}|$, L and L_z occur such that for central potentials the S-transformation is simple. A convenient choice are the coordinates used in [2, 5.3.4] with $\vec{x}$ and $\vec{p}$ exchanged: (see Fig. 6)

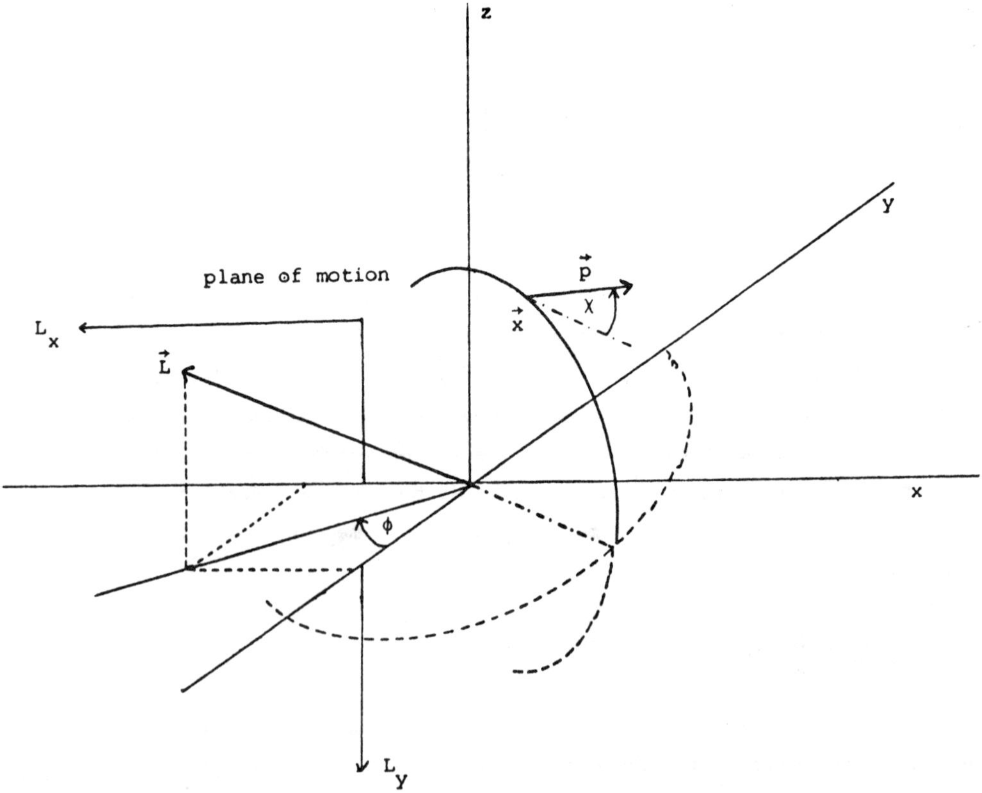

Fig. 6. The canonical coordinates in 3 dimensions

$$p = \sqrt{p_x^2 + p_y^2 + p_z^2} \ , \quad L = |\vec{x} \wedge \vec{p}| \ , \quad L_z = [x \wedge p]_z \ ,$$

$$a = \frac{\vec{x}\cdot\vec{p}}{p} \ , \quad \chi = \arc\cos \frac{L_x p_y - L_y p_x}{p\sqrt{L_x^2+L_y^2}} \ , \quad \phi = \arc\cos \frac{L_y}{\sqrt{L_x^2+L_y^2}} \ .$$

χ is the angle of momentum in the plane of motion and ϕ is the angle of the projection of $\vec{L}$ in the (x,y)-plane. For the proof of the canonicity of these coordinates see (2). The free motion is $(a,\chi,\phi;p,L,L_z) \to$ $\to (a + pt,\chi,\phi;p,L,L_z)$. For a central potential the trajectory remains in a plane and the S-transformation can be found as in b)

$$(a,\chi,\phi;p,L,L_z) \to (a - 2\,\frac{\partial\delta(p,L)}{\partial p},\ \chi - 2\,\frac{\partial\delta(p,L)}{\partial L},\phi;p,L,L_z)$$

$$\delta(p,L) = \lim_{R\to\infty} \left(\int_{r_0}^{R} dr\ \sqrt{p^2 - L^2/r^2 - 2V(r)} - \int_{r'_0}^{R} dr\ \sqrt{p^2 - L^2/r^2}\right) \ .$$

Remarks

1. We see that the generator of S is the so-called [8] quasiclassical approximation for the quantum mechanical phase shift $\delta/\hbar$. A close relation is to be expected since the quantum mechanical S-matrix $S = e^{2i\delta/\hbar}$ generates the above transformation. However the expression (6) is classically not an uncontrollable approximation but the exact result.
2. The limit $R \to \infty$ exists if V decreases for $r \to \infty$ as $r^{-1-\varepsilon}$, $\varepsilon > 0$. In this case one sees easily that also the limit $t \to \infty$ in the definition of S exists.
3. In the two- and three-dimensional case the time delay is measured by comparing the time of the trajectory with the following free motions: Follow the one tangent for $t \to -\infty$ up to r'_0, i.e. the point of closest approach to the origin, then switch over to the free trajectory tangent for $t \to +\infty$ at the same r'_0.
4. Since $pa = (xp)$ generates dilations its change under S is given by a generalization of the virial theorem to infinite trajectories. One finds for the time delay [4]

$$\tau = \frac{1}{E} \int_{-\infty}^{\infty} dt(2V(x(t)) + \vec{x}(t)\ \vec{\nabla}\ V(x(t))) \ .$$

5. Since $\delta(p,L)$ goes to zero for $L \to \infty$ we may write

$$\delta(p,L) = \frac{1}{2} \int_L^\infty dL'(\chi'(p,L,\chi) - \chi) \quad .$$

δ has the dimension of an action. Choosing $\hbar$ as unit of angular momentum

$$\frac{1}{\hbar} \delta = \frac{1}{2} \int \frac{dL'}{\hbar} (\qquad)$$

becomes dimensionless. If the deflection angle exceeds π over an interval $\hbar$ and otherwise keeps the same sign the $\delta/\hbar$ goes beyond 90°. Thus we see the following connection between resonances and looping trajectories. A looping for all angular momenta in the interval $(L', L'+\hbar)$ and $L' > L$ implies a "resonance" in the sense that the quasiclassical $\delta(p,L)$ is larger than 90°. Generally resonances occur for those L for which the sum of the deflection angles for larger L's reaches 180°. An analogous statement can be made with respect to the time delay since

$$\dot{\delta}(p,L) = \int_p^\infty p' \, dp' \, \tau(p',L) \quad .$$

6. If $H = H^0 + \lambda V$ we see

$$\frac{\partial \delta}{\delta \lambda} = - \frac{1}{2} \int_{-\infty}^{\infty} dt \, V(x(t)) \quad .$$

This is the classical version of an analogue to the Hellmann-Feynman formula in scattering theory [10]. Thus also classically $V \gtrless 0$ implies $\delta \lessgtr 0$. If we were to confine the system in a ball with radius R the right hand side above is $- R/p \cdot$(time average of V). This is a classical analogue of Schwinger's relation between phase shift and energy shift of the system in ball. This furthermore shows that Kato's monotonicity [13] is classically obvious.

7. Although we don't have a general explicit expression for $\Omega_\pm$ outside the range of the potential for $a > 0$ Ω_+ is $\mathbb{1}$ and Ω_- therefore equals S. Similarly for $a > 0$ $\Omega_- = \mathbb{1}$ and $\Omega_+ = S^{-1}$.

8. Everybody conversant with modern classical mechanics has more powerful methods available than the pedestrian ones employed sofar. He would generalize our results as follows. Since the time evolution changes the canonical 1-form by the exterior derivative of the action (see [1], (3.2.9)) we have with the previous notation

$$\vec{p}_-\cdot d\vec{q}_- = \vec{p}\cdot d\vec{q} + dw_-^0, \quad (q_+,p_-) = \Phi^0_{-t/2}(q,p), \quad w_-^0 = -\int_{-t/2}^{0} dt'\ \Phi^0_{t'}\left(\frac{\vec{p}^2}{2}\right),$$

$$\vec{p}_+\cdot d\vec{q}_+ = \vec{p}_-\cdot d\vec{q}_- + dw, \quad (q_+,p_+) = \Phi_t(q_-,p_-) \quad w = \int_0^t dt'\ \Phi^0_{t'}\left(\frac{\vec{p}_-^2}{2} - V(x_-)\right),$$

$$\vec{p}_s\cdot d\vec{q}_s = \vec{p}_+\cdot d\vec{q}_+ + dw_+^0, \quad (q_s,p_s) = \Phi^0_{-t/2}(q_+,p_+), \quad w_t^0 = -\int_{-t/2}^{0} dt'\Phi^0_{t'}\left(\frac{\vec{p}_+^2}{2}\right).$$

Adding these equations up gives

$$\vec{p}_s\cdot d\vec{q}_s = \vec{p}\cdot d\vec{q} + d(w + w_-^0 + w_+^0)$$

and thus for $t \to \infty$ the general form of the generator of S: $(q,p) \to \to (q_s,p_s)$. For the evaluation of the path integrals one may use

$$\int^t dt'\ \Phi_{t'}\left(\frac{p^2}{2} - V(q)\right) = \int_q^{q_t} dq'\ \sqrt{p^2+2V(q)-2V(q')} - Et$$

and rederive the previous expressions. Jajima [12] has shown that the difference between the action and the free action is the classical limit of the quantum phaseshift. To make it useful as generator of a classical canonical transformation is a matter of finding the appropriate canonical variables.

3. The Classical Levinson Theorem

Levinson's theorem relates the change of the phase between E = 0 and E = ∞ to the number of bound states. This seemingly wave-mechanical statement corresponds to a classical geometrical fact relating the volume in phase space of the bound orbits to the integral over the time delay. We shall now give a simple derivation of this relation using the fact that Ω and S as canonical transformation preserve the volume in phase space.

Let the phase space be decomposed by the characteristic functions χ_b and χ_s into the regions of bound orbits and scattering trajectories. χ_b is 1 in the former and 0 in the latter region and χ_s vice versa. Furthermore we first confine the integration to compact regions in phase space by the characteristic function

$$\Theta_E(x,p) = \begin{cases} 1 & \text{if } H(x,p) < E \\ 0 & \text{otherwise} \end{cases} ,$$

$$\Theta_R(x,p) = \begin{cases} 1 & \text{if } x^2 < R^2 \\ 0 & \text{otherwise} \end{cases} .$$

Now

$$\int d^\nu x\, d^\nu p\, \Theta_E\, \Theta_R = \int d^\nu x\, d^\nu p\, \Theta_E\, \Theta_R(\chi_b + \chi_s) = \tag{8}$$
$$= \int d^\nu x\, d^\nu p\, \Theta_E\, \Theta_R\, \chi_s + \int d^\nu x\, d^\nu p\, \Theta_E \circ \Omega_+\, \Theta_R \circ \Omega_+$$

since $\chi_s \circ \Omega_+$ is one. Furthermore $\Theta_E \circ \Omega_+ = \Theta(E - p^2/2)$ such that

$$\int d^\nu x\, d^\nu p\, \Theta_E\, \Theta_R\, \chi_b = \int d^\nu x\, d^\nu p\, \Theta_R[\Theta_E - \Theta(E - p^2/2)] + \tag{9}$$
$$+ \int d^\nu x\, d^\nu p\, \Theta(E - p^2/2)\, [\Theta_R - \Theta_R \circ \Omega_+] .$$

We now have to consider the limits R → ∞, E → ∞ and assume that the potential is reasonable enough so that this can be done with impunity. For the discussion of the right hand side we distinguish between different dimensions:

1) $\underline{\nu = 1}$

Assume that V(x) is uniformly bounded, $c_1 \leq V(x) \leq c_2$. Then

$$\lim_{R\to\infty} \int dx\, dp\, \Theta(R^2 - x^2)\, [\Theta(E - \frac{p^2}{2} - V(x)) - \Theta(E - \frac{p^2}{2})] =$$

$$= \int dx\; 2(\sqrt{2(E - V(x))} - \sqrt{2E}) \qquad \text{for } E > c_2 \quad . \tag{10}$$

This integral exists if V is integrable (which is necessary for the existence of scattering theory) and tends to zero for $E \to \infty$ as $-\frac{2}{\sqrt{2E}} \int dx\, V(x)$.
For discussing the last term we assume first that the potential has finite support such that $V(x) = 0$ for $|x| > R_o$. Then according to remark 7 in § 2

$$\chi_R \circ \Omega_+ = \Theta(x\, p)\; \Theta_R + \Theta(-x\, p)\; \Theta_R \circ S^{-1} \quad . \tag{11}$$

If the potential does not have finite support but decreases faster than $1/|x|^{1+\varepsilon}$ then (11) can be replaced by

$$\lim_{R\to\infty} [\Theta_R \circ \Omega_+ - \Theta(x\, p)\; \Theta_R - \Theta(-x\, p)\; \Theta_R \circ S^{-1}] = 0 \tag{12}$$

The effect of S^{-1} on x is known to be the shift $p\,\tau(p)$. The last term contributes only for $xp < 0$ and then only if an $|x| < R$ is shifted by S to be $> R$ or vice versa. Thus the x integral picks up $p\tau$ from one end or the other:

$$\lim_{R\to\infty} \int dx\, dp\, \Theta(E - \frac{p^2}{2})\, [\Theta_R - \Theta_R \circ \Omega_+] = - \int dp\, |p|\, \tau(p)\, \Theta(E - \frac{p^2}{2}) =$$

$$= - 2 \int d\varepsilon\; \tau(\varepsilon)\; \Theta(E - \varepsilon) \quad .$$

Thus finally in the limit $E \to \infty$ we arrive at

$$\int dx\, dp\; \chi_b = - \int dp\; |p|\; \tau(p) \quad . \tag{13}$$

Examples:

1. If $V(x) < 0$ then

$$\int dx\, dp\, \chi_b = \int dx\, dp\, \Theta(2|V(x)| - p^2) = \int dx\, 2\sqrt{2|V(x)|} \quad .$$

On the other hand

$$- \int dp\, |p|\, \tau(p) = - \int dx \int |p|\, dp \left(\frac{1}{\sqrt{p^2 - 2V(x)}} - \frac{1}{\sqrt{p^2}}\right) = 2 \int dx\, \sqrt{2|V(x)|} \quad .$$

2. Repulsive square well: $\chi_b = 0$. According to § 2 we have

$$\tau = 2R \begin{cases} \dfrac{1}{\sqrt{p^2-2V}} - \dfrac{1}{\sqrt{p^2}} & \text{for} \quad p^2 > 2V \\ -\dfrac{1}{|p|} & \text{for} \quad p^2 < 2V \end{cases}$$

and in fact

$$\int_0^\infty dp\, p\, \tau(p) = \int_0^{\sqrt{2V}} dp(-2R) + \int_{\sqrt{2V}}^\infty dp\, 2R\left(\frac{1}{\sqrt{p^2-2V}} - \frac{1}{\sqrt{p^2}}\right) = 0 \quad .$$

2) $\underline{\nu = 2}$

For $E > c_2$ the first term in (9) becomes simple and E-independent:

$$\int d^2x\, d^2p\, [\Theta(E - \frac{p^2}{2} - V(x)) - \Theta(E - \frac{p^2}{2})] = -2\pi \int d^2x\, V(x) \quad .$$

For calculating the last integral (11) remains valid for potentials with compact support. The statement (12) becomes too weak for two dimensions, because the x-integration runs over a sphere and there it has to be replaced by the condition

$$\lim_{R\to\infty} R^{1+\varepsilon} [\Theta_R \circ \Omega_+ - \Theta_R \circ S^{-1}\, \Theta(-x\, p) - \Theta_R\, \Theta(x\, p)] = 0 \quad . \qquad (12')$$

This condition is met by potentials decreasing faster than $1/r^{2+\varepsilon}$.

Fixing $\vec{p}$ the integration over x runs over half of the sphere after replacing Ω by S^{-1}. The time delay depends on p and L and ϕ (which we can forget after fixing $\vec{p}$). Since we already assumed that V vanishes sufficiently at infinity it follows that

$$\lim_{L\to\infty} L^{1+\delta}\ \tau(p,L,\phi) = 0$$

so that τ is integrable in L. Now we turn to the coordinates introduced in § 2 for $\nu = 2$. Since $|\vec{x}| = a + O(1/a)$ the action of S for large $|\vec{x}|$ becomes $S(|\vec{x}|) = S(|\vec{x}| - p\tau)$. Thus for $R \to \infty$ $\Theta_R - \Theta_R \circ S^{-1}$ becomes $\Theta(R - |\vec{x}|) - \Theta(R - |\vec{x}| - p\tau)$. Since the coordinates are canonical the volume element in phase space is dp da dL dϕ and the integral can be treated as for $\nu = 1$. We find

$$\lim_{R\to\infty} \int d^2x\, d^2p\, [\Theta_R - \Theta_R \circ S^{-1}]\ \Theta(-x\,p)\ \Theta(E - \frac{p^2}{2}) =$$

$$= - \int dp\, d\phi\ \Theta(E - \frac{p^2}{2})\ dL\ p\ \tau(p,L,\phi)$$

or in the case of a spherical symmetric potential

$$= - 2\pi \int d\varepsilon\ \Theta(E - \varepsilon)\ dL\ \tau(\varepsilon,L) \quad .$$

It should be noted that for $E > c_2$ the first and second expression in (9) become E-independent, thus it follows that for $E > c_2$ we must have

$$\int dL\, d\phi\ \tau(p,L,\phi) \equiv 0 \quad .$$

For spherical symmetric potentials this can be shown explicitly

$$\tau(\varepsilon,L) = 2\int dr\ [\frac{1}{\sqrt{2(\varepsilon - V(r)) - L^2/r^2}}\ \Theta(r - r_o(L,p)) - \frac{1}{\sqrt{2\ \varepsilon - L^2/r^2}}\ \Theta(r - L/p)]$$

where $V(r_o) + L^2/2r_o^2 = \varepsilon$. Thus after changing the order of integration the two contributions cancel upon L-integration:

$$\int dr\, r \left[\int_0^{r\sqrt{2(\varepsilon - V(r))}} \frac{dL}{\sqrt{2\,(\varepsilon - V(r))r^2 - L^2}} - \int_0^{r\sqrt{2\varepsilon}} \frac{dL}{\sqrt{2\,\varepsilon r^2 - L^2}} \right] = 0 \quad .$$

Thus in two dimensions Levinson's theorem reads

$$\int d^2p\, d^2x\, \chi_b = -\, 2\pi \int d^2x\, V(x) - \int d^2p\, dL\, \tau(\vec{p},L) \quad .$$

We note a correction term which has already been found in [5].

3) <u>$\nu = 3$</u>

The first term on the r.h.s. of (9) becomes now (again for $E > c_2$)

$$\int d^3x\, d^3p\, [\Theta(E - \frac{p^2}{2} - V(x)) - \Theta(E - \frac{p^2}{2})] =$$

$$= \frac{4\pi}{3}\, 2^{3/2} \int d^3x\, ([E - V(x)]^{3/2} - (E)^{3/2})$$

which in leading order in E

$$= -4\pi \int d^3x\, \sqrt{2E}\, V(x) \quad .$$

The evaluation of the last term is completely analogous as for two dimensions. The condition on the potential has to be strengthened to $V(r)$ decreases as $1/r^{3+\varepsilon}$. Thus

$$\lim_{R\to\infty} R^{2+\varepsilon}\, [\Theta \circ \Omega_+ - \Theta_R \circ S^{-1}\, \Theta(-x\, p) - \Theta_R\, \Theta(x\, p)] = 0$$

suffices for replacing Ω by S. The integration over the surface of the halfball can be replaced in the limit $R \to \infty$ by that over the halfphase such that we obtain

$$\lim_{R\to\infty} \int d^3x\, d^3p\, [\Theta_R - \Theta_R \circ S^{-1}]\, \Theta(-x\, p)\, \Theta(E - \frac{p^2}{2}) =$$

$$= - \int dp\, d\phi\, d\chi\, \Theta(E - \frac{p^2}{2})\, dL\, dL_z\, p\, \tau(p,L,L_z,\phi,\chi) \quad .$$

Generally the result for $\nu = 3$ is

$$\int d^3p\, d^3x\, \chi_b = \lim_{E\to\infty}[-\sqrt{2E}\, 4\pi \int d^3x\, V(x) - \int_0^{\sqrt{2E}} dp \int d\phi\, d\chi\, dL_z\, dL\, p\tau] \quad .$$

Since the left hand side of (9) becomes independent of E for sufficiently large E we obtain the relation

$$\int d\phi\, d\chi\, dL_z dL\tau = 4\pi \int d^3x\, [\sqrt{2(E - V(x))} - \sqrt{2E}] \quad \forall\, E > \sup_x V(x)\ .$$

For a spherical potential this relation can be checked explicitely since

$$\int d\phi\, d\chi\, dL_z\, dL\tau(p,L) = 4\pi^2 \int_0^{L_{max}} dL^2\, 2\int dr[\frac{1}{\sqrt{2(E-V(r))-L^2/r^2}} - \frac{1}{\sqrt{2E-L^2/r^2}}]$$

$$= 16\pi^2 \int r^2 dr[\sqrt{2(E-V(r))} - \sqrt{2E}]$$

It should be noted that in more than 1 dimension a negative potential does not necessarily generate a negative time delay. Though the particle becomes faster it may have to cover a longer trajectory.

4. The Coulomb Potential

For $V = e^2/r$ $\lim_{t\to\infty} \Phi_t \circ \Phi^0_{-t}$ does not exist: Φ^0_{-t} maps $\vec{x}$ into $\vec{x}-\vec{p}t$

but for the Kepler motion this quantity goes for $t \to \infty$ as lnt. Following Dollard [7] one considers another Φ^0_t which has also straight trajectories but covered with a non uniform speed. In the notation of §2 in two and three dimensions the free motion Φ^0_{-t} is

$$a \to a - pt - \frac{e^2}{p^2} \ln(t+1) \frac{p}{2}$$

the other variables remaining constant. Then the chain of maps (5) changes a (for $t\to\infty$) into

$$a - pt - \frac{2e^2}{p^2} \ln tp/4 + \int_{a_-}^{a_+} d\alpha \qquad \text{where}$$

$$t/2 = \int_{r_0}^{R} \frac{dr}{\sqrt{p^2 - \frac{2e^2}{r} - \frac{L^2}{r^2}}} = \frac{1}{p}\sqrt{\left(R - \frac{e^2}{p^2}\right)^2 - \frac{L^2}{p^2} - \frac{e^4}{p^4}} +$$

$$+ \frac{e^2}{p^3} \operatorname{arcosh} \frac{R - e^2/p^2}{\sqrt{\frac{L^2}{p^2} + \frac{e^2}{p^4}}} \overset{R\to\infty}{\simeq} \frac{R}{p} - \frac{e^2}{p^3} + \frac{e^2 \ln R}{p^3} -$$

$$- \frac{e^2}{2p^3} \ln\left(L^2 + \frac{e^4}{p^2}\right) + \frac{e^2}{p^3} \ln p/2$$

and $a_{\pm} = R + O(1/R)$. Thus for $R\to\infty$ we find

$$S(a) = a + \frac{2e^2}{p^2} + \frac{e^2}{p^2} \ln\left(L^2 + \frac{e^4}{p^2}\right).$$

Using the well known expression for the Coulomb scattering angle we have in three dimension.

$$S(a,\chi,\phi,p,L,L_z) \to (a - 2\frac{\partial\delta(p,L)}{p}, -2\frac{\partial\delta(p,L)}{L}, \Phi; p,L,L_z)$$

$$\delta(p,L) = \frac{1}{2i}\left[(L + \frac{ie^2}{p})\ln(L + \frac{ie^2}{p}) - (L - \frac{ie^2}{p})\ln(L - \frac{ie^2}{p})\right]$$

Remarks

1. Φ_t^o is for fixed t a canonical transformation but Φ^o is not a one parameter group. Its choice is to a large extend arbitrary, one has only to see that the ln R term cancels. This liberty affects the time delay but not the scattering angle.
2. It is remarkable that for the smooth potentials $V = 1/r$ or $1/r^2$ the classical and quantum δ's are so similar: Essentially one has to replace L by $1/2 + \sqrt{\vec{L}^2 + 1/2}$ to obtain the quantum phase shift. In the Coulomb case we have chosen Φ^o such that there is no additional term depending only on p. Such a contribution only enters into the time delay and depends on the choice of Φ^o. Φ determines δ up to a function of p only. Also quantum mechanically the Coulomb phase shift can be deduced up to a function of p by studying the asymptotic properties of Φ [11].
3. The change in Φ^o does not repair Levinson's theorem for the Coulomb potential. Also the analog of Hellmann-Feynman's theorem and therefore the sign rule are not valid in this case. Note that now for $e^2 > 0$ (repulsive case) we have $\tau < 0$ whereas for $V = e^2 r^{-\nu}$, $\nu > 1$, we get according to remark 6 of § 2 that

$$\tau = \frac{e^2}{E}(2-\nu)\int_{-\infty}^{\infty} dt\ r(t)^{-\nu} .$$

References

[1] W. Hunziker: Scattering in Classical Mechanics, in: Scattering Theory in Mathematical Physics, J.A. Levita and J. Marchands eds, Boston: D. Reidel 1974

[2] W. Thirring, Classical Dynamical Systems, Springer, New York (1978)

M. Breitenecker, W. Thirring: Suppl. Nuovo Cim. 2/4,p1 (1979)

[3] E. Wigner, Phys. Rev. 98, 145 (1955)

[4] H. Narnhofer: Another Definition of Time Delay, to be published in Phys. Rev.

[5] T.A. Osborn, R.G. Froese, S.F. Howes: Phys. Rev. A22, 101 (1980)

D. Bollé, T.A. Osborn: Sum Rules in Chemical Scattering, Preprint KUL Leuven

[6] W. Thirring: Quantenmechanik von Atomen und Molekülen p. 156, Springer, Wien (1979), and

H. Narnhofer, W. Thirring: The Canonical Scattering Transformation in Classical Mechanics, Vienna preprint 80-25, submitted to Phys. Rev.

[7] J.D. Dollard: J. Math. Phys. 5, 729 (1964)

[8] M. Reed, B. Simon: Scattering Theory p.11 Academic Press, New York (1979)

[9] L.D. Landau, E.M. Lifschitz: Vol. III, Quantummechanics § 126, Pergamon, Exford (1958)

[10] R. Blankenbekler, R. Sugar, Phys. Rev. 136, 472 (1964)

[11] H. Grosse, H.R. Grümm, H. Narnhofer, W. Thirring: Acta Phys. Austr. 40, 97 (1974)

[12] K. Jajima, in: Mathematical Problems in Theoretical Physics, p. 73, Springer Lecture Notes in Physics 116

[13] T. Kato, Hadronic Journal 1, 1, 134 (1978)

II . DIFFERENTIAL OPERATORS ON MANIFOLDS AND QUANTIZATION METHODS

COADJOINT ORBITS AND A NEW SYMBOL CALCULUS FOR LINE BUNDLES

Bertram Kostant
Massachussetts Institute of Technology
Cambridge, U.S.A.

The symbol calculus for differential operators is the inverse of quantization. That is, let

$$\partial = \sum_{|\epsilon| \leq k} a_\epsilon(u) \left(\frac{\partial}{\partial u}\right)^\epsilon$$

be a differential operator of degree k on $\mathbb{R}^m$. Here $\epsilon = (i_1, \cdots, i_m) \in \mathbb{N}^m$ is a multi-index, $|\epsilon| = \sum_{j=1}^m i_j$ and $\left(\frac{\partial}{\partial u}\right)^\epsilon = \left(\frac{\partial}{\partial u_1}\right)^{i_1} \cdots\cdots \left(\frac{\partial}{\partial u_m}\right)^{i_m}$. Then the symbol $\sigma_k(\partial)$ is the function on phase space $\mathbb{R}^{2m}$ given by

$$\sigma_k(\partial) = \sum_{|\epsilon| = k} a_\epsilon(q)\, p^\epsilon$$

One must consider only the terms where $|\epsilon| = k$ in order to make correspondence $\partial \to \sigma_k(\partial)$ invariant under canonical transformations and also to make commutation go to Poisson bracket. However in ignoring terms where $|\epsilon| < k$ there is clearly a great loss of information. For scalar valued functions $\varphi(u)$ this loss can be dealt with in many cases (e.g. the elliptic theory). On the other hand the symbol calculus is valid for line bundles (or even vector valued) functions $\varphi(u)$. Here the loss of information can be very serious indeed. An important example of this is the case of coadjoint orbits of Lie groups.

In more detail let G be a connected Lie group, $\mathfrak{g}$ its Lie algebra and $\mathfrak{g}'$ the dual space to $\mathfrak{g}$. One knows that any orbit $O \subseteq \mathfrak{g}'$ of G (that is, a coadjoint orbit) has the structure of a symplectic manifold (that is, a generalized phase space) which is invariant under G. Assume that O is prequantizable and that there is a G-invariant polarization P. Let $S(\mathfrak{g})$ and $U(\mathfrak{g})$ be respectively the symmetric and enveloping algebras of $\mathfrak{g}$. Then there is a line bundle L(O/P) and a representation

$$\pi : U(\mathfrak{g}) \longrightarrow \mathrm{Diff}\,(O/P,\ L(O/P))$$

of $U(\mathfrak{g})$ as differential operators on L(O/P) (i.e. an induced representation). The operators $\pi(U(\mathfrak{g}))$ can be regarded as a quantization of the functions $S(\mathfrak{J})|0$, obtained by restricting $S(\mathfrak{J})$ to 0. The failure of the symbol calculus is manifest in that one does not obtain $S(\mathfrak{J})|0$ by applying the symbol calculus to $\pi(U(\mathfrak{g}))$. In fact 0 does not appear, in general, in the cotangent bundle $T^*(O/P)$.

The purpose of this note is to announce a new symbol calculus for line bundles which overcomes the difficulty mentioned above. Let (L, M) be a pair where M is any manifold and L is a Hermitian line bundle over M. Let $m = \dim M$. We introduce what we refer to as the L-shifted cotangent bundle $X = X(L, M)$ of M. This was introduced independently and earlier by Weinstein (in generalizing some results of Sternberg) for other purposes having to do with connections in L. One has a fibration

$$\tau : X \to M$$

where the fibres are the leaves of the polarization. X is defined simply by taking the cotangent bundle of L and then reducing by the $\mathbb{C}^*$ action.

Now one has a surjective map

$$\mathrm{Diff}^* L \to \mathrm{Diff}(M, L)$$

where Diff(M, L) are the differential operators on M operating on sections of L and Diff*L are ordinary differential operators on L which commute with the $\mathbb{C}^*$ action. The point, however, is that any ∂ in Diff(M, L) only has a symbol on $T^*(M)$ whereas any $\partial \in$ Diff*L has a symbol on X. The latter is much sharper. For example if ∂ is a first order operator then can be trivialized as $\partial = \sum_{i=1}^{m} a_i \frac{\partial}{\partial u_i} + b$.
The usual symbol ignores b. But for the new symbol

$$\sigma_R(M, L) : \mathrm{Diff}_R^* L \to C^\infty(X)$$

one has $\sigma_1(M, L)\partial = \sum a_i(q) P_i + b(q)$.
The reason why this works is that the multiplication operator b is actually a vertical vector field on L. Applied to groups this recovers the coadjoint orbit theory.

One has now the following theorem, characterizing the subgroups of a Lie group which polarize coadjoint orbits.

Theorem. Let G be any connected Lie group and let (H, χ) be any pair where H is a closed subgroup of G and $\chi\ H \to \mathbb{C}^*$ is a unitary character on H. Let L(G/H) be the line bundle over G/H defined by putting

$$L(G/H) = G \times_H \mathbb{C}_\chi$$

Then H is a polarization at some $f \in \mathfrak{g}$ where $\mathfrak{g}$ is the Lie algebra of G, and $\chi = \exp 2\pi i f$ on H_e if and only if there exists an open orbit of G in the shifted cotangent bundle $X(G/H, L(G/H))$.

Furthermore we can solve a problem on the kernel of certain induced modules of $U(\mathfrak{g})$. Let $f \in \mathfrak{g}'$ and $\eta \leq \mathfrak{g}$ be a polarization at f so that $\mathbb{C}$ becomes a $U(\eta)$ module. Let

$$V = U(\mathfrak{g}) \otimes_{U(\eta)} \mathbb{C}$$

so that V is an induced $U(\mathfrak{g})$ module. Let $I \leq U(\mathfrak{g})$ be the kernel of the representation of $U(\mathfrak{g})$ on V. Let $\mathrm{Gr}\, I \leq S(\mathfrak{g})$ be the associated graded ideal in $S(\mathfrak{g})$. Let

$$\mu : T^*(G/H) \to \mathfrak{g}'$$

be the usual moment map on the cotangent bundle $T^*(G/H)$.

Theorem. Assume that G is algebraic and O, the orbit through f, is closed, or more generally assume that $\bar{O}$ is a normal variety. Then $\mathrm{Gr}\, I$ is the ideal in $S(\mathfrak{g})$ which vanishes on the image $\mu(T^*(G/H))$.

Remark. In the semi-simple case if O is closed the image of μ is the closure of the Richardson orbit defined by the parabolic η. The theorem above solves a problem raised, I believe, by W. Borho.

KÄHLER STRUCTURES FOR SOLUTION VARIETIES OF PSEUDODIFFERENTIAL OPERATORS

Stig I. Andersson

Institut für Theoretische Physik
Technical University of Clausthal
Federal Rep. Germany

ABSTRACT: We propose and study an a priori characterization of the solution varieties of non-linear pseudo-differential operators in terms of Kähler geometry. Besides being of intrinsic interest for the theory of non-linear partial differential operators, this description has direct applications in quantization theory. Additionally, within this framework interesting questions in cohomology theory as well as in the theory of unbounded operator algebras naturally emerge.

INTRODUCTION

The main objective of our study will be an a priori characterization of the solution varieties of non-linear (pseudo-) differential operators in terms of Kähler geometry. The basic ingredients are the construction, for linear pseudodifferential operators (psdo) of a certain skew-symmetric operator using the theory of Fourier integral operators (fio), combined with results on linearization stability for a large class of psdo.

The motivations for this study, could be summarized as follows,

(a) in dealing with non-linear partial differential operators (pdo) or more generally psdo, one has as a rule no control globally over the solution "manifold". We shall seek here to gain a priori global information about these "manifolds". This information will be expressed in terms of Kähler geometry and hence no information outside that category is included.

(b) for finite dimensional manifolds, one has a very tight connection between the geometry of a (compact) manifold and the (elliptic) psdo operating on the manifold. This connection is typically produced via the asymptotic expansions of the form

$$\operatorname{trace}(e^{-tP}) \underset{t \longrightarrow 0+}{\sim} (4\pi t)^{-n/2} \sum_{\mu=0}^{\infty} a_\mu t^\mu$$

where P denotes an elliptic psdo, n=dimension of the manifold and $\{a_\mu\}$ are directly related to the geometry/topology of the manifold.

More recent work of Connes[3] establishes such a link for geometric and operator theoretic properties in terms of the type of a certain v. Neumann algebra of pdo:s on the leaves of a foliated manifold on the one hand, and the geometry of the foliation on the other hand. Our work here should be viewed in a similar vein!

(c) applications include;

i) quantization theory of non-linear pdo:s amounting to the construction of so-called quantization functors.

ii) instanton/soliton problems. It seems that a key to the understanding of the structure of non-linear pdo:s in general lies in tracing out certain algebraic infinite dimensional submanifolds among their solutions. There are indications, that these submanifolds are the integral manifolds of the almost Kähler structures dealt with here!*

(d) within the framework presented here, a number of interesting questions in infinite dimensional (co-)homology theory and algebraic topology (including the homotopy invariant functor $X \to Ext(X)$) as well as in abstract operator theory arise very naturally.

The structure brought forth in this presentation is a fairly mixed one and we shall therefore have to be contented with a sketch of the main ingredients. We shall proceed mainly via examples and describe the general structure in a formal way. A detailed exposition of these ideas in the context of propagation of singularities for non-linear operators will appear in the Proc. of the Summer School on Non-linear Partial Differential Operators and Quantization Procedures (Clausthal, 1981).

1. Linear Partial Differential and Pseudodifferential Operators

Let $P(D) \in \mathrm{Diff}^{\mu}(\mathbb{R}^n)$ (pdo:s of order μ on $\mathbb{R}^n$) with constant coefficients and of real principal type;

$p_\mu(\beta)$ real, $p_\mu'(\beta) \neq 0$ on $\mathbb{R}_n \setminus 0$ (p_μ = principal symbol and $\beta \in \mathbb{R}_n$ denotes the cotangent variable). We are interested in a particular kind of fundamental solutions. Denoting by p the

*This was pointed out to the author in a letter from professor Yu. I. Manin.

total symbol of $P(D)$, we first simplify by assuming $p(\beta)$ real and $p'(\beta) \neq 0$ on the characteristic set $N_p = \{\beta \in \mathbb{R}_n : p(\beta) = 0\}$. To say E is a tempered fundamental solution is to say $P(D)E = \delta$, $E \in \rho'(R^n)$ i.e. $\hat{E}(\beta)$ solves the division problem $p(\beta)\hat{E}(\beta) = 1$. Now, $p'(\beta) \neq 0$ on N_p implies that the following composition makes sense;

$$\hat{E}^{\pm} := (p \pm i0)^{-1} = (t \pm i0)^{-1} \circ p = \lim_{\mu \longrightarrow 0} (p \pm i\mu)^{-1} \text{ in } \rho' .$$

Going out into the complex by means of a suitable vector field $v(\beta)$ we obtain via the Cauchy integral formula the following representation:

$$E^{\pm}(f) = \int_{a=\pm\varepsilon v(\beta)} \hat{f}(-z)p(z)^{-1}dz_1 \wedge \dots \wedge dz_n \quad \text{with} \quad z_i = \beta_i \pm ia_i \; .$$

$E^{\pm}$ are the fundamental forward/backward tempered solutions. An extra argument shows that

$$u \longrightarrow E^{\pm} * u \quad \text{is a continuous map} \quad H^s_c \longrightarrow H^{s+m-1}_{loc} \; .$$

Again, by composition and from $(t+i0)^{-1} - (t-i0)^{-1} = -2\pi i\delta$ we get $S := (E^+ - E^-) = -2\pi i\delta(p)$. A partition of unity and a local coordinate change

$$\begin{cases} y_1 = p(\beta) \\ y_2 = \beta_2 \\ \vdots \\ y_n = \beta_n \end{cases}$$

yields

$$(S*f,f)_{L^2} = -i(2\pi)^{1-n} \int_{N_p} |f(\beta)|^2 \frac{dv}{|p'(\beta)|}$$

With respect to $(\cdot,\cdot)_{L^2}$ we have $S^* = -S$ and we make the following observations;

i) S projects onto $\operatorname{Ker} P(D)$ and has a <u>large kernel</u>, in fact

$$\operatorname{Ker} S = \{u : \operatorname{supp}\hat{u} \cap N_p = \emptyset\}$$

ii) assuming that we study $P(D)$ as acting on H (=some H^s as a rule), equipping H with a new scalar product $(.,.) = \operatorname{Re}(.,.)_{L^2}$ then $w(f,g) = (Sf,g)$ is a degenerate skew-symmetric 2-form.

iii) on $H/\mathrm{Ker}S$ we obtain via polar decomposition $S = |S|J$ a complex structure J, $J^2 = -1$, $J \varepsilon \mathrm{End}(H/\mathrm{Ker}S)$ and also that $h(f,g) = a(f,g) + iw(f,g)$ with $a(f,g) := (|S|f,g)$, $w(f,g) := (Sf,g)$, is an hermitean scalar product. Removing the simplifying assumption $p'(\beta) \neq 0$ on N_p the above construction will for an operator of real principal type yield parametrices, i.e. operators $E^{\pm}$ such that

$$P(D)E^{\pm} = \delta + R^{\pm}$$

where $R^{\pm}$ is a well-behaved (regularizing) operator. Later on we shall indicate that such parametrices are as useful for our purposes as a fundamental solution. Example: Let $P(D)$ be the Klein-Gordon operator, i.e. $p(\beta) = \beta_0^2 - \beta_1^2 - \ldots - \beta_n^2 - m^2$. In this case there is an unique Lorentz invariant candidate for S, given by the integral kernel $-i\,\mathrm{sgn}(\beta_0)\delta(\beta^2 - m^2)$. Adding a static perturbation so to have $P(D)u = V(x')u$, $x = (x_0, x') \varepsilon \mathbb{R}^{n+1}$, and requiring suitable boundary conditions as well as putting conditions on V renders $C^2 := (m^2 - \Delta) + V(x')$ self adjoint. Thus the dynamical problem is solved by the solution operator

$$U(t) = \begin{pmatrix} \cos tC & C^{-1}\sin tC \\ -C\sin tC & C\cos tC \end{pmatrix} = e^{tT} \;;\; T := \begin{pmatrix} 0 & 1 \\ C^2 & 0 \end{pmatrix} .$$

Polar decomposition yields $T = |T|J$, $J^2 = -1$ with $J = \begin{pmatrix} 0 & -C^{-1} \\ C & 0 \end{pmatrix}$

making $w(f,g) = (Jf,g)_{H^s \oplus H^{s-1}}$ a degenerate skew 2-form on the Cauchy data space $(H^s \oplus H^{s-1})(\mathbb{R}^n)$.

Remark 1: J is exactly the C-structure which unitarizes $U(t)$.

The structure we shall set up in some generality is modelled on these simple examples and is suitably described with the following standard terminology; given a finite dimensional manifold M, an almost Hermitean structure is a triple (a,J,w) such that $w = a \circ J$ (i.e. $w(X,Y) = a(JX,Y)$) where $a(.,.)$ is a Riemannian metric, w is an almost symplectic structure (not necessarily closed) and J is an almost complex structure (not necessarily integrable) i.e.

$$J: M \ni x \longrightarrow J_x \varepsilon \mathrm{End}(T_x(M)) \;,\; J_x^2 = -1 .$$

By a well known result, J is a complex structure iff the torsion vanishes;

$$0 = \mathrm{Tor}(J)(X,Y) := \frac{1}{4}([JX,JY] - [X,Y] - J[JX,Y] - J[X,JY]) .$$

The almost Hermitean structure (a,J,w) is an Hermitean structure iff J is a complex structure and (a,J,w) is an almost Kähler structure iff w is a symplectic structure and finally we say that the almost Hermitean structure (a,J,w) is a Kähler structure if $Tor(J) = dw = 0$ (in which case $a(.,.)$ is called the associated Kähler metric). In the situation of relevance to us, i.e. $\dim M = \infty$ there is some further complication of a functional analytic character; let M be a Banach manifold, modelled on the Banach space $\mathbb{B}$. Then $w(.,.)$ and $a(.,.)$ induce naturally maps $w^{\#}$ and $a^{\#}$ respectively, mapping $T(M) \longrightarrow T^{*}(M)$ and in the infinite dimensional case $w^{\#}, a^{\#}$ might be injective or bijective. Correspondingly we say

$w(.,.)$ is a strong (weak) (almost) symplectic structure if $w^{\#}$ is bijective (injective) and that $a(.,.)$ is a strong (weak) Riemannian metric if $a^{\#}$ is bijective (injective). It is easy to see that;

M carries a strong almost symplectic structure iff $\mathbb{B}$ is reflexive and

the canonical 2-form on M is weakly symplectic (since the canonical 2-form is intrinsic but not all B-spaces are reflexive!)

Remark 2: There is no Darboux' theorem for weakly symplectic structures (Marsden).

Now the construction made above for $P(D) \varepsilon \mathrm{Diff}_{\mu}(\mathbb{R}^n)$ with constant coefficients generalizes directly to a manifold, M say and moreover the construction can be done more generally for variable coefficient operators in $\mathrm{Diff}_{\mu}(M)$. However since this is also true for a still larger class of operators, the pseudodifferential operators, we shall briefly sketch the construction in this case.

Let $P \varepsilon \mathrm{PDiff}_{\mu}(M)$ (properly supported psdo:s of order μ and type $(1,0)$ on M). Assume P has real homogenous principal part p of order μ and that no complete bicharacteristic strip of P remains in a compact set. Then[4] to each choice of orientation C diagonal $= C^{+} \cup C^{-}$ of the bicharacteristic relation of P

$$C = \{((x_1,\beta_1),(x_2,\beta_2)) \varepsilon p^{-1}(0) \times p^{-1}(0) : (x_1,\beta_1),(x_2,\beta_2) \text{ lie on the same bicharacteristic strip}\}$$

there exist parametrics $E^{\pm}$ i.e. $PE^{\pm} \equiv 1$ (mod operators with C^{∞}-kernel) such that for the wave-fronts we have

$$WF'(E^{\pm}) = (\text{diagonal in } T_{\circ}^{*}(M)) \cup C^{\pm}$$, with the stability property

$$PE \equiv 1 \ (\text{mod } C^{\infty}) \wedge WF'(E) \subset (\text{diag. } T_{\circ}^{*}(M)) \cup C^{\pm} \Longrightarrow E \equiv E^{\pm} \ (\text{mod } C^{\infty}).$$

Furthermore, as in the constant coefficient case above:

$E^{\pm}: H^s_c \longrightarrow H^{s+m-1}_{loc}$ continuously.

Defining $S:=(E^+ - E^-)$ for a given orientation, gives $S \varepsilon I^{1/2-m}(M\times M, C')$ (fio of order 1/2-m) and non-characteristic at every point of C' ("'" indicates multiplication with -1 in the fibre variables).

Theorem: Assuming $P^* = P$ then S could be chosen skew symmetric and $i^{-1}S$ is a non-negative operator on $C^\infty_\circ(M)$.

Sketch of proof: $PE^+ = 1 + R$ so $(E^+)^*P = 1 + R^*$; $WF'(E^+) = \text{diag}.\ T^*_\circ(M) \cup C^+$ implies $WF'((E^+)^*) = \text{diag}.\ T^*_\circ(M) \cup C^-$ since C^+ and C^- are inverse relations. Hence $(E^+)^*$ is a left parametrix and by stability property $(E^+)^* \equiv E^-$ (mode C^∞) , i.e. $((E^+)^*-E^-) + ((E^+)^*-E^-)^* = S^* + S \equiv 0 \pmod{C^\infty}$ Thus $S^* \equiv -S \pmod{C^\infty}$. Positivity is much harder and is achieved through reduction microlocally to $D_1 = \frac{\partial}{\partial x}$.

Remark 3: The situation is highly unstable and it is not sufficient to require self adjointness to a certain order.

Now to conclude the linear case, by appropriate conditions at infinity (in particular for compact manifolds) one can construct exact skew symmetric operators S with the above properties[5]. This is exactly what is expected on heuristic grounds viz. that the situation gets fixed only under suitable boundary conditons!

It is now easy to see that in following the pattern for the constant coefficient operator above and starting from the skew symmetric S attached to a given $P \varepsilon \text{PDiff}_\mu(M)$, we can in fact construct an almost Hermitean structure (a_P, J_P, w_P) It is however also fairly simple to see that this almost Hermitean structure is in fact a Kähler structure for any linear psdo, i.e. $\text{Tor}(J_P) = dw_P = 0$. So to describe $H':=H/\text{Ker}S$ in terms of Kähler geometry doesn't yield any interesting information in the linear case. In the non-linear situation, this set-up will prove highly non-trivial though!

Remark 4: The almost Hermitean structure (a_P, J_P, w_P) is natural to P and enjoys a certain stability property in addition.

2. Non-linear Operators, Linearization Stability

In this case $\text{Tor}(J)$ and dw become crucial global quantities. Basically the procedure is the following one; set up the above structure for the linearized operator i.e. locally in each tangent space. Then under the assumption of linearization stability, we shall try and patch these local structures together. This will in general yield a global almost Hermitean structure since in general the almost complex structure will not be integrable nor will the almost symplectic (in fact weak symplectic) structure be closed. This depends directly on the character of the non-linear operator P, giving the link between operator properties and geometry in this infinite dimensional setup!

Without entering a detailed account, we shall slightly more precisely study the following situation:

Let $P: X \longrightarrow Y$ be a non-linear psdo, X,Y Frechet manifolds (C^{∞} -sections of some vector bundle). Fixing $g_\circ \varepsilon Y$, $Pf = g_\circ$ will be a system of non-linear partial differential equations. The linearized system around the solution $f_\circ$ to $Pf=g_\circ$ is the system

$$TP(f_\circ)h = 0 , \quad h \varepsilon T_{f_\circ} X$$

i.e. the solutions of the linearized system are $\mathrm{Ker}\, TP(f_\circ)$. Assuming for simplicity that $TP(f_\circ)$ is a scalar operator and a psdo of real principal type (which can be described in terms of direct conditions on P!). Then the above construction for the linear case, produces an almost Hermitean structure

$$(a_{f_\circ}, J_{f_\circ}, w_{f_\circ})$$

defined on $\mathrm{Ker} TP(f_\circ)$.

Next piece of information needed is that of linearization stability;

$Pf = g_\circ$ is linearization stable at $f_\circ$ if $\forall h \varepsilon \mathrm{Ker} TP(f_\circ)$ there is a curve f_t for small t of solutions $Pf_t = g_\circ$ such that $f_{t=0} = f_\circ$ and $f'_{t=0} = h$.

Criterion for linearization stability at $f_\circ$: Assume X,Y are Banach-manifolds and $P:X \longrightarrow Y$ is a C^1 -map and $Pf_\circ = g_\circ$ for a fixed $g_\circ$. Suppose $TP(f_\circ)$ is a submersion (surjective and splitting kernel) then $Pf = g_\circ$ is linearization stable around $f_\circ$.

Proof: $\mathcal{M} = P^{-1}(g_\circ)$ is locally around $f_\circ$ a C^1-submanifold with $T_{f_\circ} \mathcal{M} = \mathrm{Ker} TP(f_\circ)$. Thus for any $h \varepsilon \mathrm{Ker} TP(f_\circ)$ there exist a curve $f_t \varepsilon \mathcal{M}$ and tangent to h, by definition of a manifold.

$Pf = g_\circ$ is linearization stable if it is linearization stable for all f, implying that

$$\mathcal{M}_{g_\circ} := \{f \varepsilon X: Pf = g_\circ\}$$

has a well defined tangent space everywhere $T_f \mathcal{M}_{g_\circ} = \mathrm{Ker} TP(f)$.

Remark 5: More refined criteria for linearization stability is obtained by using transversality.

Assuming now linearization stability and $TP(f_\circ)\varepsilon PDiff_\mu$ and of real principal type we can construct on $T_{f_\circ}\mathfrak{m}_{g_\circ}$ parametrices $E^{\pm}(f_\circ)$ and $S(f_\circ) := E^{+}(f_\circ) - E^{-}(f_\circ)$ such that

$$TP(f_\circ)S(f_\circ) = 0 \ (\text{mod } C^\infty)$$
$$TP(f_\circ)E^{\pm}(f_\circ) = 1 \ (\text{mod } C^\infty) .$$

Following the pattern of the linear case, we obtain on $T_f\mathfrak{m}_{g_\circ}$ $\forall f$ an almost Hermitean structure (a_f, J_f, w_f) .

The global topology/differential geometry enters the picture when we try and patch these local structures together to get a global almost Hermitean structure. As a result of this patching process, we shall obtain for linearization stable operators, a priori information about $\mathfrak{m}_{g_\circ}$: $g_\circ\varepsilon Y$ fixed in terms of the above Kähler data, by investigating closeness of w and integrability for J in the global almost Hermitean structure. To make this more precise we would have had to give a detailed account of the constructions sketched above and which is out of question here. We refer to the article in the forthcoming proceedings from the Clausthal Summer School for the details of this framework in a slightly different context[1].

Remark 6: From the global behaviour of dw a modest starting point for obtaining information about connectivity properties of $\mathfrak{m}_{g_\circ}$ has been reached.

3. Applications

Having given a brief account of the basic steps in the a priori global characterization of the solution manifolds of non-linear, linearization stable psdo:s, let me also indicate some of the possible applications.

I. Quantization of non-linear pdo:s

For an introduction to these questions cf. Refs.(2) and (6). By quantization we mean basically a (covariant) functor from the classical category (symplectic manifolds, symplec-tomorphisms) to the quantum category (Hilbert spaces, unitary operators) fulfilling some physical requirements. In particular a Weyl-quantization is a strongly ray-continuous multiplier representation of the symplectic manifold made up of $\mathfrak{m}_g$ (almost symplectic, weakly symplectic etc.) into the Weyl algebra, where the multiplier is given by the symplectic form i.e. the only data from the "classical" non-linear pdo which survive and in fact characterize the quantized version of the system is its character of an almost Hermitean structure (Hermitean structure, almost Kähler or Kähler). Through the procedure described above, we can a priori characterize the solution manifold of a non-linear pdo (psdo) in exactly those terms relevant for quantization theory!

II. The homotopy invariant functor $X \longrightarrow \mathrm{Ext}(X)$

In this setting of quantization theory, via the so called Shale criterion, the question of characterizing abstractly operators with the property of being isometric modulo the Hilbert-Schmidt operators ($\mathfrak{S}_2$) arises;

$$T^*T - 1 = 0 \ (\mathrm{mod}\ \mathfrak{S}_2) .$$

This leads naturally to an extension of the Brown/Douglas/Fillmore theory for classifying almost normal (isometric) operators, the solution of which turns out to be deeply connected to algebraic topology, expressed through the extensions. From the operator theoretic point of view, the problem is connected to finding a substitute for the index and essential spectrum (connected to the Calkin algebra, i.e. connected with invariance under compact perturbations). This substitute should have the same properties relative to $\mathfrak{S}_p$ (Schatten-von Neumann classes) as index and essential spectrum have relative to $\mathfrak{S}_\infty$ (compact operators).

References

1. Stig I. Andersson: "Kähler Geometry and Propagation of Singularities for Non-linear Partial Differential Operators", Proc. from the Summer School on Non-Linear Partial Differential Operators and Quantization Procedures, Clausthal 1981.
2. Stig I. Andersson: "Unitary Implementation of Second Quantized Dynamics of Hyperbolic Type", Rep. Math. Phys. 9, 393-422 (1976).
3. Alain Connes: "Sur la théorie non-commutative de l'intégration" in Lecture Notes in Mathematics, volume 725, pp. 19-143.
4. J.J. Duistermaat, L. Hörmander: "Fourier Integral Operators II", Acta Math. 128, 183-269 (1972).
5. J.J. Duistermaat, J. Sjöstrand: "A Global Construction for Pseudo-Differential Operators with Non-Involutive Characteristics", Inventiones Math. 20, 209-225 (1973).
6. I.E. Segal: "Symplectic Structures and the Quantization Problem for Wave Equations", Symp. Math. XIV, 99-117 (1974).

GROUP ACTION ON HOMOGENEOUS SPACES AND SOLUTIONS OF NONLINEAR DYNAMICAL SYSTEMS

A.O. BARUT

Department of Physics, The University of Colorado

Boulder, Colorado 80309

Abstract: To every nonlinear group action on a homogeneous corresponds a nonlinear integrable dynamical system.

1. INTRODUCTION

This is an elementary introduction to a method of solution of some classes of nonlinear ordinary differential equations by recognizing that the equations of the nonlinear dynamical system are the infinitesimal form of a nonlinear group realization in some homogeneous space.

Nonlinear dynamical systems exhibit, as the parameters of the system are varied, widely differing behaviour such as periodic limit cycles, nonperiodic oscillations, chaotic behaviour, strange attractors. When the parameters take certain special symmetric values, the system correspopds to a nonlinear group realization so that the boundaries of these different regions may be characterized by a dynamical symmetry group.

2. LINEAR ACTION AND LINEAR DYNAMICAL SYSTEMS

We begin, in order to illustrate the method, by the simple linear action of a group G on a space V.

Let $\Lambda(a)$ be a representation of G, $\{a\}$ the set of group parameters. In a coordinate representation we have

$$\eta_i(a)=\Lambda_{ij}(a)\eta^j(0) \quad ;i,j = 1,2,\ldots,N+k. \qquad (1)$$

We suppose that the parameters $\{a\}$ themselves depend on another parameter t, then

$$\frac{\partial \eta_i(a)}{\partial t} = \frac{\partial \Lambda_{ij}(a)}{\partial a_k} \dot{a}_k \eta^j(0) \tag{2}$$

or, using the inverse of (1), $\eta_j(0) = (\Lambda^{-1})_{jk} \eta^k(a)$, we get the equations of a dynamical system

$$\dot{\eta}_i = \frac{\partial \Lambda_{ij}}{\partial a_k} \dot{a}_k \Lambda^{-1\,j\ell} \eta_\ell \equiv L_{i\ell} \eta^\ell . \tag{3}$$

Let $\Lambda(a) = \exp(a^i L_i)$, where L_i are the infinitesimal generators of G. In this form we cannot easily differentiate $\Lambda(a)$ with respect to a_i and t because of noncommutativity of L_i. However, if we have a group action along a particular one-parameter subgroup with generator L_k , then

$$\Lambda(a) = \exp(a_k L_k) \qquad \text{(no sum)}$$

then eq.(3) becomes

$$\dot{\eta}_i(a) = (L^k)_{ij} \dot{a}_k \eta^j(a). \tag{4}$$

(sum over j but no sum over k).

This is the equation of a linear dynamical system whose solution is given by eq.(1).

Example : Let G be SU(2) or SO(3) and $L_k = \begin{pmatrix} 0 & 1 \\ -1 & 0 \end{pmatrix}$.

The dynamical system is

$$\begin{aligned} \dot{\eta}_1 &= \dot{a} \eta_2 \\ \dot{\eta}_2 &= -\dot{a} \eta_1 \end{aligned} \tag{5}$$

and the solution is clearly

$$\begin{pmatrix} \eta_1 \\ \eta_2 \end{pmatrix} = \begin{pmatrix} \cos a & \sin a \\ -\sin a & \cos a \end{pmatrix} \begin{pmatrix} \eta_1(0) \\ \eta_2(0) \end{pmatrix} = e^{aL} \vec{\eta}(0),$$

where a(t) is any differentiable function of t.

3. NONLINEAR GROUP ACTION

We shall now derive the dynamical systems corresponding to a nonlinear action of G an a space G/H, where H is a closed subgroup of G. We begin with some simple specific cases.

(i) The Projective Group

Let again

$$\eta_\alpha(a) = \Lambda_{\alpha\beta}(a)\eta^\beta(0)\,; \quad \alpha,\beta = 1\ldots(N+1) \tag{6}$$

be a linear action of G in (N+1)-dimensions.

We define the homogeneous coordinates y by

$$y_i \equiv \eta_i/\eta_{N+1} \quad , y_{N+1} = 1 \tag{7}$$

Inserting the linear transformations of η_i we have

$$y_i(a) = \frac{\Lambda_{ij}\,\eta^j(0) + \Lambda_{iN+1}\,\eta^{N+1}(0)}{\Lambda_{N+1j}\,\eta^j(0) + \Lambda_{N+1,N+1}\,\eta^{N+1}(0)} \quad , i = 1,\ldots,N ,$$

which can be expressed as a transformation of the coordinates y

$$y_i(a) = \frac{\Lambda_{ij}\,y^j(0) + \Lambda_{iN+1}}{\Lambda_{N+1j}\,y^j(0) + \Lambda_{N+1,N+1}} \,. \tag{8}$$

We recognize the nonlinear projective realization of G in N-dimensions in which the N-dimensional subgroup of G is in fact linearly represented.

We shall now find the dynamical system corresponding to this nonlinear action. For simplicity we consider a particular one-parametersubgroup so that for the linear action we have

$$\dot{\eta}_\alpha = \frac{\partial \Lambda}{\partial a}\,\dot{a}\,\Lambda^{-1}\eta = \dot{a}\,L\eta \,. \tag{9}$$

Consequently, from (7),

$$\dot{y}_i = [\dot{\eta}_i\eta_{N+1} - \dot{\eta}_{N+1}\eta_i]/\eta^2_{N+1} \,.$$

We use now (9) on the right hand side and reexpress everything in terms of $\{y\}$ and obtain

$$\dot{y}_i = \dot{a}\left[L_{ik}y^k + L_{iN+1} - L_{N+1k}y^k y_i - L_{N+1,N+1}y_i\right]. \tag{10}$$

Dynamical systems of this type characterize the projective representations. They are coupled systems of Riccati equations with a symmetry of coefficients of the following type :

$$\dot{y}_i = A_i + (B_{ik} + B\delta_{ik})y^k + C_k y^k y_i \,. \tag{11}$$

The solution of the nonlinear system (10) is therefore given by (8) where

$$\Lambda(a) = \exp aL .$$

Thus, given a system of the type (11), identify the coefficients A, B, C with the matrix elements of L according to eq.(10), then write down with the help of (13) the solution (8).

Example : The symmetric Gause - Lotke - Volterra system

$$\begin{aligned} \dot{y}_1 &= y_1 - y_1^2 - \beta y_1 y_2 - \gamma y_1 y_3 \\ \dot{y}_2 &= y_2 - \gamma y_1 y_2 - y_2^2 - \beta y_1 y_3 \\ \dot{y}_3 &= y_3 - \beta y_1 y_3 - \gamma y_2 y_3 - y_3^2 , \quad \text{for } \beta = \gamma = 1 , \end{aligned} \tag{12}$$

is of the form (11) and its solution can be obtained immediately by the method described to be

$$y_i(t) = e^t y_i(0) \left[(e^t - 1) \sum_k y_k(0) + 1 \right]^{-1} . \tag{13}$$

(11) Nonlinear Action of the Conformal Group

We start again from the linear action of G, in this case in an (N+2)-dimensional space :

$$\eta_\alpha(a) = \Lambda_{\alpha\beta}(a)\, \eta^\beta(0) \; ; \alpha, \beta = 1, \cdots, N+2 ,$$

and define

$$\begin{aligned} y_i &= \eta_i / \eta_{N+1} , \quad i = 1, \ldots, N \\ y_{N+1} &= 1 \\ y_{N+2} &= \eta_{N+2} / \eta_{N+1} . \end{aligned} \tag{14}$$

Consider now the cône in the (N+2)-dimensional space

$$\eta_i \eta^i - \eta_{N+1} \eta_{N+2} = 0 \tag{15}$$

which defines an (N+1)-dimensional homogeneous space. The equation of the cône in terms of the y-coordinates is

$$y_i y^i (\eta_{N+1})^2 - \eta_{N+1} \eta_{N+2} = 0 ,$$

or, for $\eta_{N+1} \neq 0$,

$$y_i y^i \equiv \vec{y}^{\,2} = y_{N+2} . \tag{16}$$

Thus y_{N+1} and y_{N+2} are determined and we have a nonlinear realization in the N-dimensional space of y_i which can be written as

$$y_i(a) = \frac{\Lambda_{ij}\, y^j(0) + \Lambda_{iN+1} + \Lambda_{iN+2}\, \vec{y}^{\,2}(0)}{\Lambda_{N+1j}\, y^j(0) + \Lambda_{N+1,N+1} + \Lambda_{N+1,N+2}\, \vec{y}^{\,2}(0)} \qquad (17)$$

To find the corresponding dynamical system we proceed in the same way as in the projective case. First we differentiate eq. (14) and use the linear evolution to obtain

$$\dot{y}_i = \dot{a}\left[\frac{L_{ij}\eta^j + L_{iN+1}\eta_{N+1} + L_{iN+2}\eta_{N+2}}{\eta_{N+1}} - \frac{(L_{N+1j}\eta^j + L_{N+1,N+1}\eta^{N+1} + L_{N+1,N+2}\eta^{N+2})\,\eta_i}{\eta^2_{N+1}}\right] \qquad (18)$$

or using the constraints in (14) and (15) we get the final form of the dynamical system

$$\dot{y}_i = \dot{a}\left[L_{ij}\, y^j + L_{iN+1} + L_{iN+2}\, \vec{y}^{\,2} - L_{N+1j}\, y^j y_i - L_{N+1,N+1}\, y_i - L_{N+1,N+2}\, y_i \vec{y}^{\,2}\right] \qquad (19)$$

There is however a consistency condition. Note that the elements of L : $L_{N+2,i}$, $L_{N+2,N+1}$ and $L_{N+2.N+2}$ do not appear in (19) and the corresponding group transformations in (17). They have to be chosen in such way that the cône-condition (15) remains true under the evolution. More precisely we must have transitive action of G on the cône. We have two other evolution equations for $\eta_{N+1}(a)$ and $\eta_{N+2}(a)$ in terms of $\eta_j(0)$, hence

$$y_{N+2}(a) = \frac{\Lambda_{N+2j}\, y^j(0) + \Lambda_{N+2N+1} + \Lambda_{N+2,N+2}\, y_{N+2}(0)}{\Lambda_{N+1j}\, y^j(0) + \Lambda_{N+1,N+1} + \Lambda_{N+1,N+1}\, y_{N+2}(0)}$$

The numerator of this equation must be so chosen that

$$y_{N+2}(a) = \vec{y}^{\,2}(a)$$

for all a , where the right hand side is calculated from (17).

The conformal transitive group action on the cône is well-known if G is the orthogonal or pseudoorthogonal group, in particular the conformal group in Minkowski space, for which the special conformal transformations are

$$y_i(a) = \frac{y_i(0) + a_i \vec{y}^{\,2}}{1 + 2\vec{a}\cdot\vec{y} + a^2 y^2}. \tag{20}$$

In this case we can go one step further, and consider the action of G on an (N-1)-dimensional space, thus, for example, a realization of the conformal group of the Minkowski space on a three-dimensional space. This is obtained by observing that

$$\vec{y}^{\,2}(a) = \vec{y}^{\,2}(0)\,\sigma(y) \tag{21}$$
$$\sigma(y) = (1 + 2\vec{a}\cdot\vec{y} + a^2 y^2)^{-1}.$$

Thus, the "light-cone"

$$\vec{y}^{\,2} = 0 \tag{22}$$

is invariant under the action of G. On this submanifold, we can eliminate one of the coordinates of y, for example y_N :

$$y_N = r = \pm \left[\sum_{1}^{N-1} y_K y^K \right]^{1/2}. \tag{23}$$

The action of the conformal group G_{N+2} on this (N-1)-dimensional space is given by

$$y_s(a) = \frac{\Lambda_{sr} y^r + \Lambda_{sN} r}{1 + 2(a_r y^r + a_N r)}, \quad s, r = 1, \dots, N-1 \tag{24}$$
$$y_N(a) = r.$$

The corresponding nonlinear dynamical system is then

$$\dot{y}_m = L_{mn} y^n + L_{mN} r - L_{N+1,n} y^n y_m - L_{N+1,N} r y_m - L_{N+1,N+1} y_m. \tag{25}$$

(iii) <u>Nonlinear Realization of O(N=1) on S^N - -Model</u>

The (N+1)-dimensional linear realization of G is given by

$$\eta_\alpha(a) = O_{\alpha\beta}\,\eta^\beta(0), \quad \alpha, \beta = 1, \dots, N+1.$$

The sphere S^N is defined by the invariant

$$\eta_\alpha(a)\eta^\alpha(a) = R^2 = \eta_\alpha(0)\eta^\alpha(0).$$

Hence we can eliminate the coordinate η_{N+1}, for example, and obtain

$$\eta_i(a) = O_{ij}\,\eta^j(0) + O_{i\,N+1}(R^2 - \vec{\eta}^{\,2})^{1/2} \tag{27}$$
$$\eta_{N+1}(a) = O_{N+1\,j}\,\eta^j(0) + O_{N+1,N+1}(R^2 - \vec{\eta}^{\,2}(0))^{1/2}$$
$$i, j = 1, \dots, N; \quad \vec{\eta}^{\,2} = \eta_i \eta^i$$

The corresponding nonlinear differential equations are

$$\dot{\eta}_i = L_{ij}\eta^j + L_{iN+1}(R^2-\vec{\eta}^2)^{1/2}$$
$$\dot{\eta}_{N+1} = L_{N+1j}\eta^j + L_{N+1,N+1}(R^2-\vec{\eta}^2)^{1/2}. \qquad (28)$$

The general theory is now clear. Let H be a closed subgroup of G. On G/H the subgoup H is represented linearly and there is a nonlinear action of the remaining group elements on this space. In the example of the conformal group we have, for example,

$$G = SO(4,2)\ ,\ H = T^4 \rtimes SO(3,1) \rtimes D\ ,$$

where D is the group of dilatations, and

$$G/H \sim M^4,$$

the Minkowski space.

If we write the equations of our dynamical systems in the form

$$y_i(t) = \sum Z_k(t)\,\xi^k_i(y)$$

then the vector fields defined by

$$X^k = \xi^k_i(y)\,\partial/\partial y^i$$

generate a nonlinear realization of the Lie algebra of G. Hence according to the fundamental theorem of S.Lie the dynamical system has a set of fundamental solutions so that its general solution can be expressed in terms of these fundamental solutions and N arbitrary constants. This last problem (sometimes called "the nonlinear superposition principle") has been studied in great detail recently from the point of view of nonlinear realization of groups.[1]

I should like to thank R.L.Anderson for many stimulating discussions

References

1. R.L.Anderson, Lett. Math. Phys. 4,1 (1980)
R.L.Anderson,J.Harnad and P.Winternitz, Lett. Math.Phys. 5 , 143 (1981)

SYMPLECTIC ANALOGIES

Victor Guillemin and Shlomo Sternberg
Department of Mathematics
Harvard University, Cambridge, Mass., USA

The purpose of this paper is to lay out some schemes in the program of geometric quantization which emphasize the role of algebras with involution. This point of view was introduced by Weinstein [9] (with a thrust on the group theoretic setting) and emphasized to us by Emch (cf. [2]) in the setting of quantization and polarization. We are grateful to both for many useful discussions.

In both settings one has a symplectic manifold Y, a coisotropic submanifold $\Sigma \subset Y$ and a Lagrangian submanifold $\Gamma \subset Y \times Y \times Y^-$ which can be regarded as defining a binary composition on Lagrangian submanifolds of Y, sending Λ_1 and Λ_2 into $\Gamma \circ (\Lambda_2 \times \Lambda_1)$. The coisotrope Σ satisfies $\Gamma \circ (\Sigma \times \Sigma) \subset \Sigma$ so the set of $\Lambda \subset \Sigma$ form a "subalgebra" and, in fact, a commutative one, as will be seen. In the polarization setting, $Y = X \times X^-$ where X is a symplectic manifold, Γ is the usual composition and Σ is the "leaf relation" of a polarization on X. The details for this case will be spelled out in §3. In the group case, $Y = T^*G$, where G is a Lie group, Γ is the (twisted) normal bundle to the graph of the multiplication map and Σ is the inverse image of the origin in g^* under the moment map of the conjugation action.

In the group case the coisotropic Σ was first introduced by Kostant (unpublished) in conjunction with his study of character formulas for semisimple Lie groups. Many of the ideas in this paper are the outgrowth of extended conversation and correspondence with him and we are happy to express our indebtedness to him.

In an as yet unpublished paper of ours, "Homogeneous quantization and multiplicities of group representations," we develop a symplectic procedure for describing the K-types of representations of compact Lie groups, K, by Fourier integral operators. Among other things, the material below describes the symplectic underpinnings of this result.

1. It has been observed by several authors [4], [9] that there is a symplectic "category" whose objects are symplectic manifolds, X, Y, etc., and whose morphisms Mor(X,Y) are Lagrangian submanifolds of $X^- \times Y$ (where X^- denotes the manifold X with its symplectic form ω replaced by $-\omega$), and $X^- \times Y$ is then given the product symplectic structure. Composition is composition of relations, so if

$$\Gamma_1 \in \mathrm{Mor}(X,Y) \qquad \text{and} \qquad \Gamma_2 \in \mathrm{Mor}(Y,Z)$$

then

$$\Gamma_2 \circ \Gamma_1 \in \mathrm{Mor}(X,Z)$$

is given by

$$\Gamma_2 = \{(x,z) \mid \exists y \in Y \quad \text{with} \quad (x,y) \in \Gamma_1 \quad \text{and} \quad (y,z) \in \Gamma_2\} \; .$$

In other words, if $\pi : \Gamma_1 \times \Gamma_2 \to X^- \times Z$ is projection onto the first and last factor, $\pi(x,y_1,y_2,z) = (x,z)$, then

$$\Gamma_2 \circ \Gamma_1 = (\Gamma_1 \times \Gamma_2 \cap X + \Delta + Z)$$

where $\Delta = \{(y,y)\}$ is the diagonal in $Y \times Y$. It is a theorem, cf. [3], that $\Gamma_2 \circ \Gamma_1$ is a Lagrangian submanifold of $X^- \times Z$ provided that the intersection on the right-hand side is clean, in which case we say that we have a clean composition. This is the meaning of the quotation marks in "category" in that composition is taken to be defined in the "category" if and only if it is clean. (We prefer the quotation marks to a new word such as pseudo-category.)

It was observed by Weinstein [9] that it follows that for any symplectic manifold X, the space Mor(X,X) consisting of all Lagrangian submanifolds of $X \times X^-$ inherits the structure of a "*-algebra" in the sense that we have an associative "multiplication" (defined under the clean composition hypothesis)

$$\mathrm{Mor}(X,X) \times \mathrm{Mor}(X,X) \to \mathrm{Mor}(X,X) \quad ,$$

an identity for this multiplication (the diagonal in $X \times X$) and product reversing involution, $*$, sending $\Lambda \to \Lambda^t$ where $\Lambda^t = \{(y,x) \mid (x,y) \in \Lambda\}$. In fact, Weinstein points out that we can regard the composition as a Lagrangian submanifold

$$\Gamma \subset (X \times X^-) \times (X \times X^-) \times (X \times X^-)$$

where

$$\Gamma = \{((p,q),(q,r),(p,r))\}$$

and

$$\Gamma \circ (\Lambda_1 \times \Lambda_2) = \Lambda_2 \circ \Lambda_1 \quad .$$

In the quantization yoga, we think of a symplectic manifold X as being the classical analogue of a Hilbert space. The space X^- is the symplectic analogue of the dual space and $X_1 \times X_2$ the symplectic analogue of the (completed) tensor product space $H_1 \hat{\otimes} H_2$. Thus $X \times X^-$ is the symplectic analogue of $H \hat{\otimes} H^* \cong \mathrm{End}(H)$ and so the "*-algebra" defined above is the symplectic analogue of End(H).

Weinstein [9] also points out that there is a similar Lagrangian submanifold

$$\Gamma_G \subset T^*G \times T^*G \times T^*G$$

for any Lie group, G, which gives a $*$-algebra which is the symplectic analogue of the convolution algebra $L^2(G)$. Indeed, Γ is the (twisted) normal bundle to the graph of the multiplication map: recall [3] that if $f : M \to N$ is a smooth map, then the twisted normal bundle, to its graph, $N^{\dot{-}}(\mathrm{gr}\ f) \subset (T^*M)^- \times T^*N$, defined by

$$N^-(\text{gr } f) = \{(x, df_x^*\gamma, f(x), \gamma) \quad , \quad \gamma \in T^*N_{f(x)}\}$$

is a Lagrangian submanifold of $T^*M^- \times T^*N$. Taking $M = G \times G$, $N = G$ and $f(a,b) = ab$ gives Γ. Explicitly, if $u \in TG_a$ and $v \in TG_b$, then

$$df_{(a,b)}(u,v) = dr_b u + d\ell_a v$$

where $r_b : G \to G$ denotes right multiplication by b and $\ell_a : G \to G$ denotes left multiplication by a. Thus, for $\gamma \in T^*G_{ab}$

$$df^*_{(a,b)}\gamma = (dr_b^*\gamma, d\ell_a^*\gamma)$$

and so

$$\Gamma_G = \{(a, dr_b^*\gamma), (b, d\ell_a^*\gamma), (ab, \gamma) \mid \gamma \in T^*G_{ab}\} \quad .$$

All the above formulae can be found in Weinstein [9].

Let us now rewrite the formula for Γ_G in terms of the left invariant identification of T^*G with $G \times g^*$. Thus, at each $c \in G$, we map $d\ell_c^{*-1} : TG_e^* = g^* \to T^*G_c$ or, inversely, $d\ell_c^* : TG_c^* \to g^*$.

Thus we write

$$\gamma = d\ell_{ab}^{*-1}\xi = d\ell_a^{*-1} \circ d\ell_b^{*-1}\xi \quad .$$

Then, at the point b, the covector

$$d\ell_a^*\gamma = d\ell_b^*\xi$$

is also identified with ξ. At the point a, we have

$$\begin{aligned} dr_b^*\gamma &= dr_b^* d\ell_a^{*-1} d\ell_b^{*-1}\xi \\ &= d\ell_a^{*-1} dr_b^* d\ell_b^{*-1}\xi \end{aligned}$$

since right and left multiplication commute. But

$$dr_b^* d\ell_b^{*-1}\xi = dA^*_{b^{-1}}\xi$$

where $A_c : G \to G$ is conjugation by c, and the linear map $dA^*_{b^{-1}} : T^*G_e \to T^*G_e$ is just the coadjoint action of $b \in G$ on $\xi \in g^*$, which we shall simply write as $b\xi$. Thus $dr_b^*\gamma$ is identified with $b\xi$ and under the left invariant identification of T^*G with $G \times g^*$, the Lagrangian Γ is given by

$$\Gamma_G = \{((a,b\xi),(b,\xi),(ab,\xi) \ , \ \xi \in g^*\} \quad .$$

This expression for Γ has an interesting interpretation in terms of the moment map: We refer the reader to [6] or [8] for the basic definitions and properties of Hamiltonian group actions and the associated moment map. Let $G \times X \to X$ be a Hamiltonian action whose associated moment map is $\Phi : X \to g^*$. There is then an associated moment Lagrangian, $M \in T^*G \times X \times X^-$. In terms of the left invariant identification of T^*G with $G \times g^*$, it is given by

$$M = \{((a,\Phi(x)),x,ax)\} \quad .$$

Now left multiplication gives an action of G on itself, and this induces a Hamiltonian action of G on T^*G, which, in terms of the left invariant identification, is given by

$$a(b,\xi) = (ab,\xi) \quad .$$

Also, cf. [6], the moment map for this action is given by

$$\Phi(b,\xi) = b\xi \quad .$$

From this we see that

Γ_G <u>is the moment Lagrangian associated</u>
<u>to the left action of</u> G <u>on</u> T^*G.

In any event, Γ_G determines a "ring structure" on the space of Lagrangian submanifolds of T^*G, which is the symplectic analogue of the convolution $\circledast$ on $L^2(G)$ (which is the definition of the group ring). We denote this operator also by $\circledast$, so

$$\Lambda_1 \circledast \Lambda_2 = \Gamma \circ (\Lambda_2 \times \Lambda_1)$$

(where again, clean composition is assumed).

Let $G \times X \to X$ be a Hamiltonian G action whose moment map is Φ and whose moment Lagrangian is $M \subset T^*G \times X \times X^-$. We can think of M as "morphism" from T^*G to $X \times X^-$ "assigning" to each Langrangian submanifold $\Lambda \subset T^*G$ the Lagrangian submanifold $M \circ \Lambda \subset X \times X^-$. Explicitly,

$$M \circ \Lambda = \{(x,y) \in X \times X^- \quad \text{with} \quad y = ax \ ,$$
$$\Phi(x) = \xi \quad \text{and} \quad (a,\xi) \in \Lambda\} \ .$$

We claim that

$$(M \circ \Lambda_1) \circ (M \circ \Lambda_2) = M \circ (\Lambda_1 \circledast \Lambda_2)$$

so that M can be thought of as the symplectic analogue of the group ring morphism associated with a representation of G. Indeed,

$$M \circ \Lambda_1 = \{(x,z) \,|\, z = bx \ , \ \Phi(x) = \xi \quad \text{and} \quad (b,\xi) \in \Lambda_1\}$$

$$M \circ \Lambda_2 = \{(z,y) \,|\, y = az \ , \ \Phi(z) = \eta \quad \text{and} \quad (a,\eta) \in \Lambda_2\}$$

so

$$(M\circ\Lambda_1) \circ (M\circ\Lambda_2) = \{(x,y) \,|\, \exists z \text{ with } (x,z)\in M\circ\Lambda_1 \text{ and } (z,y)\in\Lambda_2\}$$
$$= \{(x,y) \,|\, y=abx \text{ with } \Phi(x)=\xi \text{ and } (b,\xi)\in\Lambda_1, \ (a,b\xi)\in\Lambda_2\}$$

since $\Phi(bx) = b\Phi(x)$. But this last expression is precisely $M \circ \Gamma(\Lambda_2 \times \Lambda_1) = \Lambda_1 \circledast \Lambda_2$.

2. In the quantization yoga symplectic manifolds correspond to Hilbert spaces, Lagrangian submanifolds to (generalized) lines and co-isotropic submanifolds to (generalized) subspaces. Now, given a subspace of a Hilbert space, there is an associated self-adjoint projection operator π (with $\pi^* = \pi$, $\pi^2 = \pi$), namely projection onto the

subspace. The symplectic analogue is the following, cf. [4]:

> Let V be a coisotropic submanifold of X. Define $\Lambda(V) \subset X \times X^-$ by $\Lambda = \Lambda(V) = \{(x,y) \in X \times X^-$ such that x and y lie on the same leaf of the null foliation of $V\}$.
>
> Then Λ is a Lagrangian submanifold of $X \times X^-$ and
>
> $$\Lambda \circ \Lambda = \Lambda \quad , \quad \Lambda^t = \Lambda \quad .$$
>
> Conversely, if Λ is a Lagrangian submanifold of $X \times X^-$ satisfying the above conditions, then $\Lambda \circ X = V$ is a coisotrope of X with $\Lambda(V) = \Lambda$.

For example, let $G \times X \to X$ be a Hamiltonian group action with moment map $\Phi : X \to g^*$. Let $Q \subset g^*$ be an invariant submanifold under the coadjoint action of G on g^*, and suppose that Φ intersects Q cleanly. Then, cf. [5] Theorem 4.1, $\Phi^{-1}(Q)$ is a coisotropic submanifold of X. Its symplectic analogy interpretation is as follows: Each orbit of G acting on g^* is the symplectic analogue of an irreducible representation of G. As Q is the union of orbits, it is the analogue of a collection of irreducibles. Also X is the analogue of a representation of G. So $\Phi^{-1}(Q)$ is the symplectic analogue of the subspace transforming as the various irreducibles of L.

In particular, if $Q = O$ is an orbit of G acting on g^*, then $\Phi^{-1}(O)$ is the symplectic analogue of the subspace transforming according to the "irreducible, O". Since $\Phi^{-1}(O)$ is coisotropic, it has an associated idempotent Lagrangian,

$$\Lambda(\Phi^{-1}(O)) \subset X \times X^- \quad .$$

This Lagrangian can be constructed differently as follows: Consider the moment Lagrangian, M_O, associated to the Hamiltonian action of G on $\mathcal{O}$, so $M_{\mathcal{O}} \subset T^*G \times \mathcal{O} \times \mathcal{O}^-$ is given (in terms of the left invariant identification of $T^*G \sim G \times g^*$) by

$$M_{\mathcal{O}} = \{(a,\xi,\xi,a\xi)\,\xi \in \mathcal{O}\}$$

Let $\Delta \subset \mathcal{O} \times \mathcal{O}^-$ denote the diagonal. Then

$$M_{\mathcal{O}}^t \circ \Delta = \{(a,\xi)\,|\,a\xi = \xi, \xi \in \mathcal{O}\} \overset{\text{def}}{=} \Lambda_{\mathcal{O}}$$

is a Lagrangian submanifold of T*G known as the character Lagrangian associated to $\mathcal{O}$.

Then, up to covering,

$$M \circ \Lambda_{\mathcal{O}} = \Lambda(\Phi^{-1}(\mathcal{O})) \quad .$$

More precisely, let G_ξ denote the isotropy subgroup of a point $\xi \in g^*$, i.e., $G_\xi = \{a \in G\,|\,a\xi = \xi\}$ and let G_ξ^o denote the connected component of G_ξ. Then

$$\Lambda(\Phi^{-1}(\mathcal{O})) = \{(x,ax) \quad \text{where} \quad \Phi(x) = \xi \in \mathcal{O} \quad \text{and} \quad a \in G_\xi^o\}$$

while

$$M \circ \Lambda_{\mathcal{O}} = \{(x,ax) \quad \text{where} \quad \Phi(x) = \xi \in \mathcal{O} \quad \text{and} \quad a \in G_\xi\} \quad .$$

(So, for example, if G_ξ is connected, then $\Lambda(\Phi^{-1}(\mathcal{O})) = M \circ \Lambda_{\mathcal{O}}$.) We refer the reader to [5] where these facts are proved in a more general setting with $\mathcal{O}$ replaced by any Hamiltonian G space.

The manifolds $\Lambda_{\mathcal{O}}$ are the "character Lagrangians" associated to the orbits $\mathcal{O}$ by Weinstein [1], [8] and Kostant (unpublished). They play the role of the irreducible characters, and $M \circ \Lambda_{\mathcal{O}}$ is the symplectic analogue of the description of the projector onto subspace transforming according to an irreducible as the image of the character (considered as a function on G) under the associated ring morphism from $L^2(G)$ to operators on the representation space.

Now (for compact groups) a character is a special kind of central function, i.e., a function invariant under the conjugation action of the group on itself. Let us examine what the symplectic analogue of this notion is. Let $G \times X \to X$ be any Hamiltonian action of G on a symplectic manifold, X, with moment map $\Phi : X \to g^*$. Then the symplectic analogue of the invariants is the coisotropic submanifold $\Phi^{-1}(0)$ (where the appropriate cleanness assumption is assumed, which in this case reduces to the assumption that $\Phi^{-1}(0)$ is a submanifold). We can be more explicit about this analogy: Let ω be the symplectic form of X and, for each $\eta \in g$, let η_X denote the corresponding vector field on X. Then, by the definition of the moment map

$$(2.1) \qquad i(\eta_X)\omega = d\langle\eta,\Phi\rangle$$

where $\langle\eta,\Phi\rangle$ is the ($\mathbb{R}$-valued) function given by $\langle\eta,\Phi\rangle(x) = \langle\eta,\Phi(x)\rangle$ and where $\langle,\rangle$ denotes the evaluation map between g and g^*. If $\Lambda \subset \Phi^{-1}(0)$ is a (connected) Lagrangian submanifold, then (2.1) implies that $\omega(\eta_X,\zeta) = 0$ for any vector field ζ tangent to Λ and hence η_X is tangent to Λ. If G is connected, this implies that Λ is invariant (as a set) under G. Thus

Proposition 2.1: *If G is connected, then any connected Lagrangian submanifold $\Lambda \subset \Phi^{-1}(0)$ is invariant under G.*

Conversely, suppose that Λ is invariant under G, so that η_X is tangent to Λ for each $\eta \in g$. Then (2.1) implies that $d\langle\eta,\Phi\rangle$ vanishes on any vector tangent to Λ and hence (as Λ is connected) is constant on Λ. On the other hand, by the equivariance of Φ

$$D_{\zeta_X}\langle\eta,\Phi\rangle = \langle[\zeta,\eta],\Phi\rangle$$

for any $\zeta \in g$ and hence $\langle[\zeta,\eta],\Phi\rangle = 0$ on Λ. If we make the hypothesis that $[g,g] = g$, then this implies that $\langle\eta,\Phi\rangle = 0$. Thus,

Proposition 2.2: *Let G be connected and suppose that $[g,g] = g$. Then a connected Lagrangian manifold is G invariant if and only if $\Lambda \subset \Phi^{-1}(0)$.*

Let us specialize to the case where $X = T^*G$ and we take the G action on T^*G induced from the conjugation action of G on itself. Under the left identification of T^*G with $G \times g^*$, the moment map for this action is given by

$$\Phi(c,\xi) = c\xi - \xi ,$$

of [6]. We will let $\Sigma \subset T^*G$ denote $\Phi^{-1}(0)$, so

$$\Sigma = \{(c,\xi) \mid c\xi = \xi\} \quad . \tag{2.2}$$

By Propositions 2.1 and 2.2 we have

Proposition 2.3: *Let G be a connected Lie group. Any connected Lagrangian submanifold $\Lambda \subset \Sigma$ is invariant under the action of G on T^*G induced from the conjugation action of G on itself. The converse holds if $[g,g] = g$.*

We will therefore consider the (connected) Lagrangian submanifolds of Σ as the symplectic analogue of the central functions. We would like to show that the

$$\Lambda_1 \subset \Sigma \quad \text{and} \quad \Lambda_2 \subset \Sigma \Rightarrow \Lambda_1 \circledast \Lambda_2 \subset \Sigma \quad . \tag{2.3}$$

Proof:

$$\Lambda_1 \circledast \Lambda_2 = \Gamma_G(\Lambda_2 \times \Lambda_1)$$

$$= \{(a,b\xi) \mid (b,\xi) \in \Lambda_1, (a,b\xi) \in \Lambda_2\} \quad .$$

But since $\Lambda_1 \subset \Sigma$, we have $b\xi = \xi$ and since $\Lambda_2 \subset \Sigma$, we have $a(b\xi) = \xi$ and hence $ab\xi = \xi$ so $\Lambda_1 \circledast \Lambda_2 \subset \Sigma$. Also

$$\text{if } \Lambda_1 \subset \Sigma \text{ and } \Lambda_2 \subset \Sigma, \quad \text{then} \quad \Lambda_1 \circledast \Lambda_2 = \Lambda_2 \circledast \Lambda_1 \quad . \tag{2.4}$$

Proof: By the above

$$\Lambda_1 \circledast \Lambda_2 = \{(ab,\xi) \mid (a,\xi) \in \Lambda_2 \quad \text{and} \quad (b,\xi) \in \Lambda_1\} \quad . \tag{2.5}$$

But $\Lambda_1 \circledast \Lambda_2 \subset \Sigma$ and so is invariant under the action of G. Now the action of $b \in G$ on $(c,\zeta) \in T^*G$ is given by (cf. [6])

$$b(c,\zeta) = (bcb^{-1}, b\zeta) \quad .$$

Applied to (2.4), we see that

$$\begin{aligned} \Lambda_1 \circledast \Lambda_2 &= \{(ba,\xi) \mid (a,\xi) \in \Lambda_2 \quad \text{and} \quad (b,\xi) \in \Lambda_1\} \\ &= \Lambda_2 \circledast \Lambda_1 \quad . \end{aligned}$$

For the purpose of the next section, we shall reformulate the last arguments in a more complicated, but more symplectic, form:

Let Σ be a coisotropic submanifold of a symplectic manifold, X. It has a null foliation (cf. [3]) where $v \in T\Sigma_x$ is tangent to the null foliation if and only if $w(u,v) = 0$ for all $u \in T\Sigma_x$. In particular, if $x \in \Lambda \subset \Sigma$ where Λ is Lagrangian, then $w(u,v) = 0$ for all u tangent to Λ and hence v is tangent to Λ. Hence

Proposition 2.4: *Let Σ be coisotropic and $\Lambda \subset \Sigma$ be Lagrangian. If $x \in \Lambda$, then the entire leaf of the null foliation through x is contained in Λ.*

In the case at hand, where $\Sigma = \Phi^{-1}(0)$, we know (cf. [6]) that the leaf of the null foliation through $x \in \Sigma$ is $G^o \cdot x$ where G^o is the connected component of G. As we are assuming that G is connected, this is just $G \cdot x$. Thus Proposition 3.4 implies Proposition 3.1.

Let $G \times X \to X$ be a Hamiltonian G action with moment Lagrangian M. Let $\Delta_X \subset X \times X^-$ be the diagonal. Then the character Lagrangian $\Lambda_X \subset T^*G$, as defined by Weinstein, *et al.* is given by

$$\Lambda_X = M \circ \Delta_X = \{(a,\Phi(x)) \mid ax = x\} \quad .$$

Since Φ is equivariant, $ax = x$ implies that $a\Phi(x) = \Phi(x)$, so we see that

$$\Lambda_X \subset \Sigma \quad . \tag{2.6}$$

(For the case $X = \mathcal{O}$ is an orbit in g^*, this reduces to the case considered previously.)

Observe that if $a \in G$ and $\eta \in g$, then the tangent vector to the curve $(\exp t\eta)a(\exp -t\eta)$ is (under left identification) $Ad_{a^{-1}}\eta - \eta$. Thus if $\xi \in g^*$ is orthogonal to all these vectors, then $a\xi = \xi$ so $(a,\xi) \in \Sigma$. Thus the normal bundle to the conjugacy classes are Lagrangian submanifolds contained in Σ. These play the role of the (δ functions of the) conjugacy classes. The analogue of the value of a central functions on a conjugacy class is the intersection of Λ with $N(C)$ (where C is a conjugacy class and $N(C)$ its normal bundle). The actual numerical value associated to each point of intersection is determined by data coming from prequantization and will not be discussed here.

3. Let X be a symplectic manifold. A (real) polarization on X is a foliation of X whose leaves are all Lagrangian submanifolds. A polarization, P, determines a submanifold, Σ, of $X \times X^-$, where

(3.1) $\Sigma = \{(p,q) \mid p \text{ and } q \text{ be on the same leaf of } P\}$.

In this section we shall prove the following results:

(3.2) Σ is an immersed coisotropic submanifold of $X \times X$;

(3.3) if the polarization P satisfies a topological condition (has trivial holomny and closed leaves), then Σ is an embedded submanifold.

Under the hypotheses of (3.3), we have

(3.4) if $\Lambda_1 \subset \Sigma$ and $\Lambda_2 \subset \Sigma$ are Lagrangian, then $\Lambda_1 \circ \Lambda_2 \subset \Sigma$, also $\Lambda^t \subset \Sigma$ if $\Lambda \subset \Sigma$;

(3.5) if $\Lambda_1 \subset \Sigma$ and $\Lambda_2 \subset \Sigma$, then $\Lambda_1 \circ \Lambda_1 = \Lambda_2 \circ \Lambda_1$;

(3.6) if $\Lambda_1 \circ \Lambda = \Lambda \circ \Lambda_1$ for all $\Lambda \subset \Sigma$, then $\Lambda_1 \subset \Sigma$.

The point is that the set of $\Lambda \subset \Sigma$ is the symplectic analogue of a maximal commutative subalgebra of the algebra of operators on H. (This point of view towards quantization has been employed by Emch [2].)

Proof of (3.2) and (3.3): Let P be a polarization on X, with $\dim X = 2n$. Near any point $x \in X$ we can find coordinates $x_1,\ldots,x_n$, $\xi_1,\ldots,\xi_n$ such that the symplectic form ω of X is given locally as

$$\omega = \Sigma d\xi_1 \wedge dx_i$$

and such that the leaves of P are given by

$$x_i = \text{const.} \quad .$$

We can thus use the x_i as coordinates on an n dimensional disk D_x through x transverse to the leaves of P. Let L be the leaf through x, and let γ be a curve passing through x lying in L, and joining x to some other point, y, in L. Let U be some tubular neighborhood of γ in X and $\pi : U \to L$ the projection onto L. Let D_y be a disk through y transverse to the leaves of P. We may assume that $D_x = D_x \cap U$ and $D_y = D_y \cap U$. For each $x' \in D_x$, there is a unique curve γ' lying in the leaf, L', through x' with $\pi\gamma' = \gamma$, and will intersect D_y at some point y'. The map, ϕ_γ, of $D_x \to D_y$ sending x' to y' is a diffeomorphism, and depends only on the homotopy class of γ (see [7] for all the details we are quoting from the general theory of foliations.) The (foliation of) the polarization P is said to have trivial holomony if, for any x and y on the same leaf, this map is independent of γ.

Let us choose coordinates $y_1,\ldots,y_n$, $\eta_1,\ldots,\eta_n$ near y such that

$$\omega = \Sigma d\eta_1 \wedge dy_i$$

and the leaves are given by y_i = const. We can use the y_i as local coordinates on D_y and then ϕ_γ will be given in local coordinates as

$$y_i \circ \phi_\gamma = f_i^\gamma(x_1,\ldots,x_n) \quad .$$

Now we can use $x_1,\dots,x_n$, $\xi_1,\dots,\xi_n$, $y_1,\dots,y_n$, $\eta_1,\dots,\eta_n$ as local coordinates on $X \times X^-$ near (x,y). Then the various components of Σ passing through (x,y) are given by

$$y_i - f_i^\gamma(x_1,\dots,x_n) = 0 \qquad i = 1,\dots,n \quad .$$

Letting $\{,\}_{X\times X^-}$ denote the Poisson bracket in $X \times X^-$ and $\{,\}_X$ the Poisson bracket on X, then

$$\{F_1(x,\xi)+G_1(y,\eta),F_2(x,\xi)+G_2(y,\eta)\}_{X\times X^-} = \{F_1,F_2\}_X(x,\xi)-\{G_1,G_2\}_X(y,\eta)$$

so

$$\{y_i-f_i^\gamma(x),y_j-f_j^\gamma(x)\}_{X\times X^-} = -\{y_i,y_j\}_X + \{f_i^\gamma,f_j^\gamma\}_X = 0 \quad .$$

This proves that Σ is an immersed coisotropic submanifold. There will be only one component passing through (x,y) if and only if the holomony is trivial, and Σ will be closed off as the leaves are closed. We have thus proved (3.2) and (3.3).

Let us now assume that the hypotheses of (3.3) hold. The leaf of the null foliation through (x,y) is then swept out by the commuting vector fields

$$H_{y_i} + H_{f_i}$$

where, for any function F on X, we let H_F denote the Hamiltonian vector field on X associated to F, i.e.,

$$i(H_F)\omega = dF \quad .$$

This can be given the following more global interpretation: To say that the holomony is trivial clearly implies that the normal bundle, NL, to the leaf L has a global trivilization, so we can write $NL = L \times N_L$ where N_L is a fixed vector space. The symplectic structure gives an isomorphism $NL_x \sim TL_x$ at each $x \in L$ and so each $v \in N_L$ determines a vector field on L. The coordinates $y_1,\dots,y_n$ we chose on D_y give a basis $e_1,.q.,e_n$ of N_L and the vector field corresponding to e_i is given near y by $H_{y_i|L}$ (and near x by $H_{f_i|L}$). In particular, the vector fields corresponding to $v \in N_L$ all commute and so we get an action of N_L on L. From what we just observed, se see that

(3.7) <u>the null foliation of Σ through (x,y) is the orbit of the diagonal action of N_L on (x,y).</u>

We can now prove (3.4) and (3.5) by arguments very close to those we used in Section 2.

Since Σ comes from a transitive and reflective relation on X, it satisfies $\Sigma \circ \Sigma \subset \Sigma$ and $\Sigma^t = \Sigma$ so (3.4) clearly holds.

<u>Proof of (3.5)</u>. Suppose that $(p,q) \in \Sigma$, so p and q both belong to the same leaf, L. Then there exists some $u \in N_L$ with $u \cdot p = q$ (where $u \circ p$ denotes the action of u on p described above). Now suppose $\Lambda_1 \subset \Sigma$ and $\Lambda_2 \subset \Sigma$. Then $(p,r) \subset \Lambda_1 \circ \Lambda_2$ if and only if there exists some $q \in X$ with $(p,q) \in \Lambda_1$ and $(q,r) \in \Lambda_2$. So $r \in L$ also, and $r = v \cdot q$ for some $v \in N_L$. Now by Proposition 2.4 and (3.7),

$$v \cdot (p,q) = (q^1, r) \in \Lambda_1$$

and, since $q = u \cdot p$ and $(v+u) \cdot p = r$, we see that $(q,r) \in \Lambda_2$ implies that

$$(-u) \cdot (q,r) = (p, q^1) \in \Lambda_2 \quad .$$

Thus there exists a q^1 with $(p,q^1) \in \Lambda_2$ and $(q^1,r) \in \Lambda_1$, so $(p,r) \in \Lambda_2 \circ \Lambda_1$, proving (3.5).

<u>Proof of (3.6)</u>. Let L be a leaf of P and take $\Lambda = L \times L$. Then Λ is clearly a Lagrangian submanifold and $\Lambda \subset \Sigma$. Then

$$\Lambda_1 \circ \Lambda = \{(p,q) \in \Lambda_1 \text{ with } q \in L\}$$

while

$$\Lambda \circ \Lambda_1 = \{(p,q) \in \Lambda_1 \text{ with } p \in L\} \quad .$$

The assumption $\Lambda \circ \Lambda_1 = \Lambda_1 \circ \Lambda$ says that $p \quad L \Leftrightarrow q \in L$. If this holds for all L, then $\Lambda_1 \subset \Sigma$.

REFERENCES

[1] R. Abraham and J. Marsden, Foundations of Mechanics, 2nd Ed., Addison-Wesley, 1978.

[2] G. Emch. Private communication.

[3] V. Guillemin and S. Sternberg. Geometric asymptotics, AMS, 1977.

[4] V. Guillemin and S. Sternberg. Some problems in integral geometry and some related problems in microlocal analysis. Am. J. of Math 101 (1979), 915-955.

[5] V. Guillemin and S. Sternberg. Moments and reductions, Proc. of Clausthal. Conference.

[6] D. Kazhdan, R. Kostant and S. Sternberg. Hamiltonian group actions and dynamical systems of Calozero type. Com. TAM, Vol. 31 (1978), pp. 481-507.

[7] G. Reeh. Theorie des foliations.

[8] A. Weinstein. Lectures on symplectic manifolds, AMS Reg. Conf. in Math., Series No. 29 (1976).

[9] A. Weinstein. Symplectic geometry, Bull. AMS (1981), 1-13

INVARIANT QUANTIZATION OF LINEAR TIME-DEPENDENT WAVE EQUATIONS

Stephen M. Paneitz
Department of Mathematics
University of California
Berkeley, California 94720

and

Institut für Theoretische Physik
TU Clausthal
Federal Republic of Germany

It has been known for nearly two decades that a linear field equation having a symplectic structure admits an algebraic quantization, which is unique up to C*-algebraic equivalence. The problem of determining a vacuum state is much more difficult; if the equation has time-independent coefficients, then a unique vacuum state, determined by invariance under the associated 1-parameter evolutionary group, is known to exist if the positive-energy condition is satisfied.

Hitherto, the best procedure apparently available for a similiar treatment, involving the determination of particular states on the Weyl algebra, for a prototypical time-dependent linear equation in Minkowski space, e.g.

$$\Box\varphi + m^2\varphi + V(t,\vec{x})\varphi = 0 \qquad (1)$$

has been the following: determine two (free) states $|0\rangle_-$ and $|0\rangle_+$, by ignoring the potential V at early and late times, respectively, and show that the same Fock space can accomodate $|0\rangle_-$, $|0\rangle_+$, and a unitary operator S_{qm} , which satisfies

$$S_{qm}\,|0\rangle_- = |0\rangle_+$$

and intertwines covariantly with the classical scattering operator S_{cl} . The conditions on (1) needed to assure this are basically that

$$S_{cl}^* S_{cl} = I + H$$

where H is a Hilbert-Schmidt operator.

While 'practical' and somewhat intuitive, this procedure remains lacking from a foundational standpoint. It (i) neglects the contribution of the interaction in the definition of physical particle states, i.e. does not determine a physical ground state for a quantization of (1) (where φ is replaced by an operator-valued distribution) and (ii) does not distinguish between stable and unstable systems (this distinction being represented in the case of time-independent equations by (1) where $V = 0$ and m is real and purely imaginary, respectively).

Recently we have shown that the physical vacuum state $|0\rangle$ (and scattering matrix S satisfying $S|0\rangle = |0\rangle$), for linear equations, is determined in generic cases, as the unique regular state invariant under the scattering automorphism of the algebra of observables (Weyl algebra). For e.g. equation (1), the conditions for this are that V should be nonnegative, sufficiently nonvanishing, and not too large. This condition is satisfied if (1) arises as the first-order variational equation to a classical nonlinear wave equation whose nonlinear interaction term has a nonnegative derivative (e.g. the so-called $\lambda\varphi^4$ equation) and if the nonlinear background field considered is sufficiently small. [2]

An adaptation of the relatively recent infinite-dimensional stability theory of M.G. Krein [1] plays an important role in these developments, which represent work in collaboration with I.E. Segal, and also results of the lecturer's doctoral thesis. See [2] and [4] for a summary of this work, and [3] and [5] for mathematical details. [5] also gives specific illustrations of the quantization on nonstatic space-times (involving a special case of (1)) .

References

1. J.L. Daleckii and M.G. Krein, Stability of Solutions of Differential Equations in Banach Space, Translations of Mathematical Monographs, vol. 43, Amer. Math. Soc., Providence, R.I., 1974.
2. S. Paneitz and I.E. Segal, "Quantization of wave equations and hermitian structures in partial differential varieties", Proc. Nat. Acad. Sci. USA, vol. 77, no. 12 (December 1980), pp. 6943-6947.
3. S. Paneitz, "Unitarization of symplectics and stability for causal differential equations in Hilbert space", J. Func. Anal., vol. 41 (1981), pp. 315-326.
4. _________, in Springer Math. Lecture Notes, vol. 905, 1982.
5. _________, "Essential unitarization of symplectics, and applications to field quantization", 1981, submitted for publication.

SU(n) BUNDLES OVER THE CONFIGURATION SPACE OF THREE IDENTICAL PARTICLES MOVING IN $\mathbb{R}^3$

F.J. Bloore
University of Liverpool, U.K.

Abstract: We study the system of three identical spinless particles moving in $\mathbb{R}^3$ and possessing an SU(n) gauge symmetry. Three such systems are possible, corresponding to the three non-isomorphic SU(n) bundles over $C_3(\mathbb{R}^3)$, the configuration space. We retract $C_3(\mathbb{R}^3)$ to a sub-complex which shows clearly how its homology arises. The three bundles can be realized as pull-backs of the universal bundle $S^7 \to S^4$ using three non-homotopic maps $C_3(\mathbb{R}^3) \to S^4$.

1. Introduction

In non-relativistic quantum mechanics the state of a system is represented by a ray in the Hilbert space of L^2 sections of some complex Hermitian vector bundle over the configuration space. The isomorphism class of the underlying principal bundle has a physical interpretation in terms of a superselection rule. Systems of three identical particles with an SU(n) gauge symmetry are of physical interest.

2. Configuration space $C_3(\mathbb{R}^3)$ of three identical particles in $\mathbb{R}^3$.

The configuration space of 3 distinguishable particles in $\mathbb{R}^3$ is [1]

$$\tilde{C}_3(\mathbb{R}^3) = \mathbb{R}^3 \times \mathbb{R}^3 \times \mathbb{R}^3 \ - \ \text{(diagonals)}$$

and

$$C_3(\mathbb{R}^3) = \tilde{C}_3(\mathbb{R}^3)/S_3 \quad (\equiv C_3, \text{ for short})$$

where S_3 acts by permutation. A spectral sequence calculation gives the cohomology:-

$$H^*(C_3,\mathbb{Z}) = \mathbb{Z}, 0, \mathbb{Z}_2, 0, \mathbb{Z}_3, \quad H^q(C_3,\mathbb{Z}) = 0 \text{ for } q>4.$$

It follows from the universal coefficient theorem that $H_3(C_3,\mathbb{Z}) = \mathbb{Z}_3$, $H_q(C_3,\mathbb{Z}) = 0$ for $q>3$. A generator, D of H_3 is a closed 3-sub-manifold which itself does not bound a 4-submanifold, W but $3D=\partial W$. It is useful to view C_3 as the space of all triangles in $\mathbb{R}^3$, including collinear ones. We may realize D as the 3-manifold of equilateral triangles of unit side and fixed centroid, O say. Then W is the 4-manifold of isosceles triangles of unit base, centroid O and height, $h < \sqrt{3}/2$. As h tends to $\sqrt{3}/2$, the isoceles triangles in the same plane, having bases 60° to each other all approach the same equilateral triangle so we need three copies of D to bound W. In the next section we give an explicit strong deformation retraction of C_3 to $W \cup D$. Hence this subcomplex carries all the topology of C_3.

3. The retraction of C_3 to $W \cup D$.

This has three steps:-

(i) Translate the triangles without rotation so their centroids lie at the origin, O.

(ii) Rescale, (dilate) each triangle until its shortest side (or sides) has unit length.

(iii) keeping the shortest side fixed in direction, the centroid fixed, move the vertex opposite the shortest side towards O until the triangle is isosceles.

Notice that this procedure takes all collinear triangles to isosceles collinear ones and all isosceles triangles with height $> \sqrt{3}/2$ to equilateral ones.

4. Classification of SU(n) bundles

Since $H^q(C_3, \mathbb{Z})=0$ for $q>4$, any SU(n) bundle over C_3 is classified by its second Cern class, $c_2 \varepsilon H^4(C_3, \mathbb{Z}_3)$, [2], and its structure group reduces to SU(2), [3], So we need only treat SU(2) bundles.

The Hopf fibration $\pi: S^7 \to S^4$ is a 6-universal principal SU(2) bundle. This means that if $\dim M \leq 6$ every principal SU(2) bundle over M is of the form $f^*(S^7)$ for some map $f: M \to S^4$. If f_1, f_2 are homotopic maps, then $f_1{}^*(S^7)$ and $f_2{}^*(S^7)$ are isomorphic bundles over M. We now construct three homotopically distinct maps from $W \cup D$ to S^4. These will pull-back S^7 to the three non-isomorphic SU(2) bundles over $W \cup D$ which necessarily extend to bundles over C_3.

Let I_h be the subset of $W \cup D$ of all isosceles triangles of height h, thus:- $W \cup D = \bigcup_{0 \leq h \leq \sqrt{3}/2} I_h$, with $I_{\sqrt{3}/2} = D$ the equilaterals and $I_0 = \mathbb{R}\mathbb{P}^2$, the collinears. In between, I_h, $0<h<\sqrt{3}/2$, is isomorphic to the lens space L(4,1). Thisis because the different orientations of an isosceles triangle can be uniquely specified by points in $\mathbb{R}\mathbb{P}^2/\mathbb{Z}_2$, where the action of $\mathbb{Z}_2$ on the space of rotations in $\mathbb{R}^3$ is given by flipping the triangle about its axis of symmetry, [4]. Similarly we can regard S^4 to be split into 3-spheres:- $S^4 = \bigcup_{0 \leq h \leq \sqrt{3}/2} S_h{}^3$, with $S^3{}_{\sqrt{3}/2}$ and $S_0{}^3$ as the north and south poles of S^4 respectively. Now any continuous map $f: I_{\frac{1}{2}} \to S_{\frac{1}{2}}{}^3$ extends to a continuous map $f': W \cup D \to S^4$ as follows. Since $\bigcup_{0<h<\sqrt{3}/2} I_h = I_{\frac{1}{2}} \times J$ and $\bigcup_{0<h<\sqrt{3}/2} S_h^3 = S_{\frac{1}{2}}{}^3 \times J$, $J=(0, \sqrt{3}/2)$, we put $f' = f \times id_J$ for $0<h<\sqrt{3}/2$ and let $f'(I_{\sqrt{3}/2}) = S^3{}_{\sqrt{3}/2}$ and $f'(I_0) = S_0{}^3$.

The map f has a homotopy invariant $k\varepsilon[I_{\frac{1}{2}},S_{\frac{1}{2}}^3]=H^3(I_{\frac{1}{2}},\mathbb{Z})=\mathbb{Z}$, [2] . We now show that f′ is homotopically trivial if and only if kmod3=0. To do this we split $W\cup D$ into two sub-spaces:-

$$X = \bigcup_{0\le h\le \frac{1}{2}} I_h,\quad Y = \bigcup_{\frac{1}{2}\le h\le \sqrt{3}/2} I_h$$

so that; $X\cup Y=W\cup D$ and $X\cap Y=I_{\frac{1}{2}}$. In the same way divide S^4 into northern (N) and southern (S) hemispheres, overlapping only at the equator $(S^3_{\frac{1}{2}})$ The map f′ now induces commutative diagrams involving the Mayer-Vietoris cohomology sequences of both spaces:-

$$\begin{array}{ccccccccccc}
\cdots & H^3(S^3) & \to & H^3(N)\oplus H^3(S) & \to & H^3(S^3) & \to & H^4(S^4) & \to & H^4(N)\oplus H^4(S) & \cdots \\
 & \downarrow & & \downarrow & & \downarrow f^* & & \downarrow f'^* & & \downarrow & \\
\cdots & H^3(W\cup D) & \to & H^3(X)\oplus H^3(Y) & \to & H^3(X\cap Y) & \to & H^4(W\cup D) & \to & H^4(X)\oplus H^4(Y) & \cdots
\end{array}$$

The relevant groups are ($\mathbb{Z}$ coefficients):-

$$\begin{array}{ccccccccc}
 & & 0 & \to & \mathbb{Z} & \xrightarrow{\times 1} & \mathbb{Z} & \to & 0 \\
 & & & & \downarrow \times k & & \downarrow f'^* & & \\
0 & \to & H^3(Y) & \to & \mathbb{Z} & \xrightarrow{\text{mod}\pm 3} & \mathbb{Z}_3 & \to & 0
\end{array}$$

In the second line $H^3(X)=H^4(X)=0$ since X retracts to $I_0= \mathbb{R}\ \mathbb{P}^2$. ($H^3(Y)$ is irrelevant; exactness of the second line implies $H^3(Y)=\mathbb{Z}$). Even though there is a sign ambiguity in the mod3 homomorphism, we see from commutativity that f'^* is trivial if and only if kmod3=0. If kmod3≠0 then we can obtain representatives of all three homotopy classes of maps f'^* by composing f with a general map from S^3 to S^3 of appropriate degree, this justifies the extension procedure. A suitable map $f:I_{\frac{1}{2}}\to S_{\frac{1}{2}}^3$ is obtained if we regard the lens space $I_{\frac{1}{2}}$ as a 3-ball with certain identifications on its boundary (S^2), [5] . f then takes the "interior" 3-cell identically to the 3-cell of S^3 and takes the "boundary" $(S^2/\sim)$ to the 0-cell of S^3 . An argument similar to the one above shows that this f has kmod3≠0.

5. Remarks

Quantum mechanics in the two non-trivial bundles has no conventional analogue. Sections s of vector bundles associated to these principal bundles pull-back to sections $\tilde{s}$ of the canonically trivial vector bundle over $\tilde{C}_3(\mathbb{R}^3)$. The sections $\tilde{s}$ do not carry a representation of S_3, because the non-trivial bundles do not admit flat connections.

References

1. F.J. Bloore, - configuration spaces of identical particles, Lecture Notes in Mathematics 836, 1-8. Springer-Verlag. Berlin 1980.
2. S.J. Avis and C.J. Isham - Quantum field theory and fibre bundles in a general space-time, in Recent developments in gravitation - Cargese 1978, eds. S. Deser and M. Levy, Plenum Press, London 1979.
3. C.J. Isham - Spontaneous symmetry breaking and topological charge:- Imperial College Theoretical Physics preprint 1981.
4. P.M. Rice - Free actions of $\mathbb{Z}_4$ on S^3. Duke Math. Journal 36 (1969) 749-751.
5. P.J. Hilton and S. Wylie - Homology Theory. Cambridge Univ. Press. 1962, p.223.

Acknowledgements

We wish to thank C.T.C. Wall and G.P. Scott for many helpful discussions. I. Bratley and J.M. Selig also thank S.E.R.C. for finanacial support.

DYNAMICAL QUANTIZATION

T.D. Palev
Institute for Nuclear Research, Sofia, Bulgaria

The main equations of quantum mechanics in the Heisenberg picutre are the Heisenberg equations

$$\dot{p}_i = -\frac{i}{\hbar}[p_i, H], \quad \dot{q}_i = -\frac{i}{\hbar}[q_i, H], \tag{1}$$

where H is the Hamiltonian and (the Cartesian) coordinates $q_1, \dots, q_n$ and momenta $p_1, \dots, p_n$ satisfy the canonical commutation relations(CCR's)

$$[q_i, p_j] = i\hbar\delta_{ij}, \quad [q_i, q_j] = [p_i, p_j] = 0. \tag{2}$$

From (1) and (2) one derives the Hamiltonian equations

$$\dot{p}_i = -\frac{\partial H}{\partial q_i}, \quad \dot{q}_i = \frac{\partial H}{\partial p_i}. \tag{3}$$

The eqs.(1) can also be derived from (2) and (3). It turns out, however, that the CCR's do not follow with necessity from (1) and (3). We use this circumstance in order to modify the concept of a quantization: the operators q_i and p_i are called position and momentum operators if they satisfy simultaneously the eqs.(1) and (3). This definition is physically justified,since the Heisenberg and the Hamiltonian equations have more immediate physical significance than the CCR's [1]. We call the quantization defined above a dynamical quantization, since in general q_i and p_i may depend on the dynamics, on the interaction.

As an example we study a dynamical quantization of a system of two nonrelativistic point particles, interacting via a harmonic potential. The centre-of-mass variables are quantized in a canonical way (2), whereas the internal coordinates r_1, r_2, r_3 and momenta p_1, p_2, p_3 are assumed to generate the simple Lie superalgebra sl(1,3). The latter requirement is not a necessary one, however,it appears as a natural generalization of the Lie superalgebraical properties of the canonical operators (2), which generate an orthosymplectic Lie superalgebra[2]. It is convenient to introduce the variables

$$a_k^{\xi} = \sqrt{\frac{m\omega}{2\hbar}}\, r_k + i\xi \sqrt{\frac{1}{2m\omega\hbar}}\, p_k , \tag{4}$$

which are reffered to as creation(ξ=+) and annihilation (ξ=-) operators(CAO's). Here m is the reduced mass and ω - the frequency of the oscillator. The CAO's (4) satisfy the relations

$$[\{a_i^+, a_j^-\}, a_k^+] = \delta_{jk} a_i^+ - \delta_{ij} a_k^+ ,$$

$$[\{a_i^+, a_j^-\}, a_k^-] = -\delta_{ik} a_j^- + \delta_{ij} a_k^- , \tag{5}$$

$$\{a_i^+, a_j^+\} = \{a_i^-, a_j^-\} = 0 .$$

We consider a class of representations of the CAO's, which are finite-dimensional and are labelled by one nonnegative integer p [3]. These representations are of Fock type. For any p the metric is positive definite and r_i, p_i are hermitian operators. The internal Hamiltonian commutes with the internal distance operator and,therefore,the distance between the particles is preserved in time. It can have only four different values

$$r^{(k)} = \sqrt{\frac{\hbar}{2m\omega}(3p-2k)} , \quad k=0,1,2,3 ; \quad k \leq p . \tag{6}$$

The internal energy, corresponding to a distance with a given k in (6) is

$$E^{(k)} = \frac{\hbar\omega}{2}(3p-2k). \tag{7}$$

Since the operators of the coordinates do not commute, the position of the particles cannot be localized in the space. Thus,the particles are moving as the ends of a stick with a certain length, which orientation in the space cannot be defined.

References

1. E.P.Wigner, Phys.Rev.77,711(1950).
2. A.Gantchev, T.D.Palev,J.Math.Phys.21,797(1980).
3. T.D.Palev. J.Math.Phys.21,1293(1980).

III . PARTICLE PHYSICS AND SPACE-TIME GEOMETRY

C_{2n} SPINOR GEOMETRY

P. Budinich and P. Furlan

International School for Advanced Studies (SISSA), Trieste, Italy

International Centre for Theoretical Physics, Trieste, Italy

and

Instituto di Fisica Teorica dell'Università di Trieste, Italy

1. INTRODUCTION

In 1935 E. Cartan [1] wrote:

"Il est intéressant de remarquer que la notion de vecteur peut se déduire de celle de spineur: du moins on peut avec un spineur simple former un ν-vecteur isotrope, puis on peut définir un ν-vecteur général comme somme de ν-vecteurs isotropes, et un vecteur comme un élément commun à une famille de ν-vecteurs satifaisant à certaines conditions."*

In these years when the experimental evidences in high energy physics seem to show us the fundamental role spinors play in the elementary structure of matter, Cartan's words may acquire the flavour of a prophecy: that if we succeed in bringing to an explicit form the possibility envisaged in those words and formulate space-time-geometry in terms of pure-spinor-geometry, then we will have brought to evidence the fundamental spinorial structure not only of space-time geometry, but also of matter. We shall then

* "It is interesting to note that the notion of a vector can be deduced from that of a spinor; at least we can form from a pure spinor an isotropic ν-vector, then a general ν-vector can be defined as the sum of isotropic ν-vectors, and a vector as a common element of a family of ν-vectors which satisfy certain conditions".

also have the key for the understanding, in purely geometrical terms, the mysterious internal symmetries which seem to govern the elementary laws of matter.

We have attempted some steps in that direction and wish to summarise here what we find and seems to us not too discouraging.

The basic idea is that pure spinors are the only elementary geometrical objects in terms of which both space-time geometry and the laws of physics in space-time must be formulated.

2. C_{2n} SPINOR GEOMETRY

Following criteria of maximal economy we first define:

Definition of C_{2n} Spinor Geometry (C_{2n}S.G.): the one which may be obtained from a pure spinor ξ, vector of the representation space of a given Clifford algebra C_{2n} (which in turn may be obtained from the exterior algebra of a 2n-dimensional Euclidean space E_{2n}, but might also be defined independently).

Its geometrical constituents are:

α) The 2^n-component pure* spinor ξ direct sum of two 2^{n-1}-component pure semi-spinors φ_I and φ_{II}.

β) The direct sum $[\varphi_I] \oplus [\varphi_{II}]$ of the two isotropic semi-n-vectors of E_{2n} homomorphic, each of them, to the pure semi-spinors φ_I and φ_{II} contained in ξ and each of them basis for inequivalent

---

* Pureness of ξ as a whole can be only defined in the frame of C_{2n+1} Clifford algebra. In this preliminary version we will, for brevity, call ξ "pure" if direct sum of two pure semi-spinors.

irreducible representations of the connected orthogonal group in E_{2n}.

γ) The isotropic intersections of the mentioned isotropic semi-n-vectors: $[\varphi_I] \cap [\varphi_{II}]$. ▼

Remark 2.1. We may consider the pure semi-spinors φ_I and φ_{II} as the elementary geometrical objects in C_{2n} S.G.. The E_{2n} tensors in β) and in γ) may be expressed as bilinear polinomials of φ_I and φ_{II} components:

β) $$V^{(n)} := [\varphi_I] \oplus [\varphi_{II}] \quad (2.1)$$

with components:

$$V^{(n)}_{[a_1 \dots a_n]} = \xi^T B \Gamma_{[a_1} \dots \Gamma_{a_n]} \xi = \varphi_{I_\alpha} M^{\alpha\beta}_{[a_1 \dots a_n]} \varphi_{I_\beta} + \varphi_{II_\alpha} N^{\alpha\beta}_{[a_1 \dots a_n]} \varphi_{II_\beta} \quad (2.1\beta)$$

where Γ_a are the generators of the Clifford algebra C_{2n} obeying:

$$[\Gamma_a, \Gamma_b]_+ = 2 g_{ab} \quad (2.2)$$

and B is a $2^n \times 2^n$ matrix with the property:

$$B \Gamma_a = (-)^n \Gamma_a^T B \quad (2.3)$$

and $M_{[\,]}$ and $N_{[\,]}$ are symmetric matrices.

γ) $$V^{(p)} := [\varphi_I] \cap [\varphi_{II}] \quad \text{for } n-p = 3 \pmod 4 \quad (2.4)$$

with components:

$$V^{(p)}_{[a_1 a_2 \dots a_p]} \quad \xi^T B \Gamma_{[a_1} \Gamma_{a_2} \dots \Gamma_{a_p]} \xi \quad \varphi_{I_\alpha} P^{\alpha\beta}_{[a_1 a_2 \dots a_p]} \varphi_{II_\beta} \quad (2.4\gamma)$$

and $P_{[\,]}$ are matrices.

Remark 2.2. There is a fundamental difference between the tensor $V^{(n)}$ in β) and $V^{(p)}$ in γ). In fact while the semi n-planes $[\varphi^{I}]$ and $[\varphi^{II}]$ in $V^{(n)}$ may be considered as equivalent [1] to φ_I and φ_{II} and as such as elementary as the semi-spinors themselves, the intersections of $[\varphi^{I}]$ and $[\varphi^{II}]$ giving $V^{(p)}$ are the only genuine "new" tensors constructed with the elementary pure spinors, in line with the Cartan programme. We shall see that the quantum theory seems to underline this difference.

Remark 2.3. The correlation between $V^{(n)}$ and the spinor ξ may be also expressed through the equation:

$$\Gamma_a P^a \xi = 0 \tag{2.5}$$

where P^a is an isotropic E_{2n}-vector: in fact from (2.5) and (2.2) we get

$$P_a P^a \xi = 0 \ . \tag{2.6}$$

Eq. (2.5) allows us to express ξ components in terms of P_a. Equation (2.5), if of rank n, defines an n-dimensional isotropic hyper-plane of E_{2n} which coincides with $V^{(n)}$. Eq. (2.5) may be expressed in terms of the elementary pure semi-spinors φ_I and φ_{II}:

$$\Gamma_a P^a (1\!\!1 + \Gamma_{2n+1})\xi = 0 \ ; \ \Gamma_a P^a (1\!\!1 - \Gamma_{2n+1})\xi = 0 \tag{2.7}$$

where

$$\Gamma_{2n+1} := i^{\alpha}\Gamma_1 \Gamma_2 \ldots \Gamma_{2n} \tag{2.8}$$

with $\alpha = n(2n-1)$ for $g_{ab} = \delta_{ab}$; and in an appropriate basis (Cartan basis):

$$\frac{1}{2}(1\!\!1 + \Gamma_{2n+1})\xi^{c} = \begin{vmatrix} \varphi_I \\ 0 \end{vmatrix} \ , \ \frac{1}{2}(1\!\!1 - \Gamma_{2n+1})\xi^{c} = \begin{vmatrix} 0 \\ \varphi_{II} \end{vmatrix} \tag{2.9}$$

We may now start by constructing the simplest C_{2n} S.G..

3. C_{2n} SPINOR GEOMETRY : $M^{3,1}$ - TENSORS AND CLASSICAL PHYSICS

3.1. C_4 S.G.

Let us suppose C_4 generated by $\gamma_0, \gamma_1, \gamma_2, \gamma_3$ Dirac matrices; the corresponding E_4 is then $M^{3,1}$ Minkowski space and the spinor ψ is then the 4-component Dirac spinor and it is pure. The tensors we may construct in $M^{3,1}$ are

$$V^{(2)} = [\varphi_I] \oplus [\varphi_{II}]$$

with components

$$\frac{1}{2} \psi^T \gamma_2 \gamma_0 [\gamma_\mu, \gamma_\nu] \psi := F_{\mu\nu} := f_{\mu\nu} + \tilde{f}_{\mu\nu} \qquad (3.1,1\beta)$$

equivalent to ψ itself and:

$$V^{(1)} = [\varphi_I] \cap [\varphi_{II}]$$

with components

$$\psi^T \gamma_2 \gamma_0 \gamma_5 \gamma_\mu \psi := V_\mu \quad . \qquad (3.1,1\gamma)$$

If we express $F_{\mu\nu}$ in terms of a vector $\bar{E}$ and a pseudovector $\bar{H}$ of E_3

$$\begin{aligned} f_{\mu\nu} &\Rightarrow \bar{H} + i\bar{E} \\ \tilde{f}_{\mu\nu} &\Rightarrow \bar{H} - i\bar{E} \quad , \end{aligned} \qquad (3.1,2)$$

the isotropy of $F_{\mu\nu}$, homomorphic and equivalent to ψ, is expressed by:

$$\begin{aligned} \bar{E} \cdot \bar{H} &= 0 \\ \bar{H}^2 &= \bar{E}^2 \quad . \end{aligned} \qquad (3.1,3)$$

$\bar{E}$ and $\bar{H}$ may then represent the electric and magnetic field of light propagating in vacuum.

The isotropy of $V^{(1)}$ reads:

$$V_\mu V^\mu = 0 \quad . \qquad (3.1,4)$$

The four-vector V_μ may then only represent momenta of massless particles or light rays.

In case of C_4S.G. the equations (2.5) and (2.7) become:

$$\gamma_\mu p^\mu \psi = 0 \quad ; \quad \gamma_\mu p^\mu (1 \pm \gamma_5) \psi = 0 \quad , \tag{3.1,5}$$

which may be formally interpreted as Dirac and Weyl equations in momentum space. We have seen that they express the correlation of the pure semi-spinors φ_I and φ_{II} with the isotropic semi-bi-vectors $[\varphi_I]$ and $[\varphi_{II}]$. From them, taking into account Eq. (3.1,1β), we may derive the Maxwell equations in momentum space:

$$p^\mu f_{\mu\nu} = 0 \quad , \quad p^\mu \tilde{f}_{\mu\nu} = 0 \quad . \tag{3.1,6}$$

This geometrical derivation of Maxwell equations from Weyl ones (or (3.1,1β)) may not however be interpreted as compositeness of the photon in terms of neutrinos, say, as we shall see.

We conclude then that C_4 spinor geometry consists of two dual isotropic bi-vectors and a vector on the light cone. In physics it is apt to represent light as propagating in vacuum, light rays, momenta of massless particles and, formally, Dirac and Maxwell equations; but neither $M^{3,1}$ vectors outside the light cone, nor massive systems may be represented in the frame of C_4S.G.. It is then necessary to go to higher dimensional spinors.

3.2. C_6S.G.

C_6 is generated by $\Gamma_0, \Gamma_1, \Gamma_2, \Gamma_3, \Gamma_5, \Gamma_6$; we will take:

$$[\Gamma_a, \Gamma_b]_+ = 2g_{ab} \; ; \; g_{aa} = (+1,-1,-1,-1,-1,+1) \quad , \tag{3.2,1}$$

and E_6 is specialized to $M^{4,2}$. C_6S.G. is constituted by a pure spinor equivalent to the direct sum of two isotropic 3-planes in $M^{4,2}$ with components:

$$V^{(3)}_{abc} = \xi^T B \Gamma_{[a}\Gamma_b\Gamma_{c]}\xi \in [\xi] = [\varphi_I] \oplus [\varphi_{II}] \quad (3.2,2)$$

and an isotropic bivector in $M^{4,2}$ with components:

$$V^{(2)}_{ab} = \xi^T B [\Gamma_a, \Gamma_b] \xi \in [\varphi_I] \cap [\varphi_{II}] \quad (3.2,3)$$

(in C_7 ξ pure satisfies equation: $\xi^T B \xi = 0$).

A (Dirac) basis may be defined in which ξ has the form:

$$\xi^D = \begin{vmatrix} \psi_2 \\ \psi_1 \end{vmatrix} , \quad (3.2,4)$$

with ψ_1 and ψ_2 Dirac spinors. In terms of these one may then "read" the $M^{3,1}$ tensor content of $V^{(3)}$ and $V^{(2)}$ (avoiding thus the difficulties inherent in dimensional reduction) and we find:

$$V^{(3)} = [\varphi_I] \oplus [\varphi_{II}] \ni A_\mu, V_\mu, F^{(1)}_{\mu\nu}, F^{(2)}_{\mu\nu} \quad (3.2,5)$$

where A_μ stands for an $M^{3,1}$ axial vector and V_μ for a $M^{3,1}$ vector and

$$V^{(2)} = [\varphi_I] \cap [\varphi_{II}] \ni P, F_{\mu\nu}, P^{(1)}_\mu, P^{(2)}_\mu \quad (3.2,6)$$

where P stands for a pseudo-scalar and P_μ for vectors (the superscripts refer to the Dirac spinor indices in (3.2,4)).

The four-vectors are not necessarily isotropic now in $M^{3,1}$ and we could also represent vectors outside the light cone (however it is not easy to construct the full geometry of $M^{3,1}$ including the generators of Poincaré group in C_6S.G., [2] this should only be possible in C_8S.G.).

The equation (2.3) which correlates the spinor ξ with $V^{(3)}$ becomes

$$P_a \Gamma^a \xi = 0 , \tag{3.2,7}$$

and is IO(4,2)-covariant. Interpreting P_a as a gradient operator in $M^{4,2}$ one may obtain from this a field equation for the spinor field $\xi(\eta)$. Introducing the variables

$$x_\mu = \frac{\eta_\mu}{k} , \quad \alpha^2 = \frac{\eta^2}{k^2} , \quad k = \eta_5 + \eta_6 \neq 0 , \tag{3.2,8}$$

equation (3.2,7) for $\xi(\eta)$ homogeneous function of η (in the Dirac basis) may be transformed into [3,4]:

$$i\gamma_\mu \frac{\partial}{\partial x_\mu} \Psi_1(x,\alpha) = 2\alpha^2 \frac{\partial}{\partial \alpha^2} \Psi_2(x,\alpha)$$

$$i\gamma_\mu \frac{\partial}{\partial x_\mu} \Psi_2(x,\alpha) = -2 \frac{\partial}{\partial \alpha^2} \Psi_1(x,\alpha) , \tag{3.2,9}$$

which is still SO(4,2)-covariant. With the ansatz

$$\Psi_1(x,\alpha) = f_1(x,\alpha)\, \psi_1(x)$$

$$\Psi_2(x,\alpha) = f_2(x,\alpha)\, \psi_2(x) \tag{3.2,10}$$

with $\psi_i(x)$ Dirac spinors, and:

$$f_1(x,\alpha) \equiv f_1(\alpha; c_1, d_1, \varphi_1(x)) = \alpha \left\{ c_1 J_1[\alpha \varphi_1(x)] + d_1 Y_1[\alpha \varphi_1(x)] \right\}$$

$$f_2(x,\alpha) \equiv f_2(\alpha; c_2, d_2, \varphi_2(x)) = c_2 J_0[\alpha \varphi_2(x)] + d_2 Y_0[\alpha \varphi_2(x)] \tag{3.2,11}$$

with $\varphi_i(x)$ α-independent and otherwise unspecified scalar functions, and c_i, d_i constants, Y_i, J_i Bessel functions, (3.2,9) may be exactly solved with respect to the α variable. In fact suppose

$$f_1(\alpha; c_1, d_1, \varphi_1) = f_1(\alpha; c_2, d_2, \varphi_2)$$

$$f_2(\alpha; c_2, d_2, \varphi_2) = f_2(\alpha; c_1, d_1, \varphi_1) , \tag{3.2,12}$$

which is satisfied if $c_1 = c_2$, $d_1 = d_2$, $\varphi_1 = \varphi_2 = \varphi$; then substituting (3.2,11) in (3.2,9), we find easily:

$$i\gamma_\mu\left(\frac{\partial}{\partial x_\mu} - i\, B^{(1)}_\mu(x)\right)\Psi_1(x) = -\,\varphi(x)\,\Psi_2(x)$$

$$i\gamma_\mu \frac{\partial}{\partial x_\mu}\Psi_2(x) = -\,\varphi(x)\,\Psi_1(x)\,, \tag{3.2,13}$$

where

$$B^{(1)}_\mu(x) = -\lim_{\alpha\to 0} i\frac{\partial}{\partial x^\mu}\ln f_1(\alpha; c_1, d_1, \varphi(x)) = -\frac{i\partial}{\partial x^\mu}\ln\varphi(x)$$

It is easily seen that, because of the known properties of Bessel function a term $B^{(2)}_\mu(x)$, analogous to $B^{(1)}_\mu$ which would appear in the second Eq. (3.2,13) vanishes in the limit $\alpha \to 0$. $B^{(1)}_\mu$ may be eliminated through a local gauge transformation on the field $\psi_1(x)$ in the first equation. But there is an even more interesting solution if we go with α to the limit $\alpha \to 0$ after substituting (3.2,10) with (3.2,11) in (3.2,9). We find the previous equations with $\varphi(x)$ substituted by $d_2/d_1\varphi_1(x)$ in the first equation and by $d_1/d_2\,\varphi_2(x)$ in the second one. If we choose the arbitrary constants d_1, d_2 to satisfy: $d_1/d_2 = e^{i\theta}$ we obtain

$$i\gamma_\mu\partial^\mu N(x) + \gamma_\mu B^{(1)\mu}(x)\frac{1-\tau_3}{2}N(x) = [\tau_1\rho_1(x) + \tau_2\rho_2(x)]\,N(x)\,,$$

where $N = \begin{vmatrix}\psi_2\\ \psi_1\end{vmatrix}$ and $\rho_1 = \varphi_1\cos\theta$; $\rho_2 = \varphi_2 \operatorname{sen}\theta$. If we now add an O(4,2)-scalar term $\phi(\eta)\,\xi(\eta)$ on the r.h.s. of Eq. (3.2,7) we arrive, for $\alpha \to 0$ at the suggestive form:

$$i\gamma_\mu\partial^\mu N + \gamma_\mu B^{(1)\mu}\frac{1-\tau_3}{2}N = \vec{\tau}\cdot\vec{\rho}\,N\,, \tag{3.2,14}$$

where $\rho_3(x) = \lim_{\alpha\to 0}\phi(x,\alpha)$.

<u>Remark 3,1</u> In general Ψ_1 and Ψ_2 in (3.2,4) and (3.2,9) correspond to different eigenvalues of the dilatation operator ($\Gamma_6\Gamma_5 = \sigma_3$ in

the Dirac basis) and consequently, as known, they may not be both canonical Dirac spinors. However since α transforms like x_μ for dilatations (see 3.2,8), $f_1(x,\alpha)$ and $f_2(x,\alpha)$ in (3.2,10) (see 3.2,11) bear the Ψ_1 and Ψ_2 difference in behaviour for dilatations; consequently $\psi_1(x)$ and $\psi_2(x)$ in Eq. (3.2,10), (3.2,13), may both represent canonical Dirac spinors.

Remark 3,2 Should we have started from the more elementary equation

$$P_a \Gamma^a (\mathbb{1} + \Gamma_7) \xi = 0 \, , \tag{3.2,15}$$

corresponding to (2.7), instead of (3.2,7), we would have obtained (taking the trivial solution of (3.2,12), remembering that Γ_7 is an SO(4,2) Casimir operator of the form $\gamma_5 \otimes \tau_3$ in the Dirac basis):

$$i\gamma_\mu \left(\frac{\partial}{\partial x_\mu} - i B^{(1)\mu}(x) \right) \psi_{1L}(x) = -\varphi(x)\, \psi_{2R}(x)$$

$$i\gamma_\mu \frac{\partial}{\partial x_\mu} \psi_{2R}(x) = -\varphi(x)\, \psi_{1L}(x) \, . \tag{3.2,16}$$

In a coherent S.G. approach $\varphi(x)$ will have to be expressed in terms of elementary spinors (say $\varphi(x) = g\, \bar{\psi}_{3_L} \gamma_\mu \psi_{4_L} \gamma^\mu$).

Equation (3.2,7) may be generalized both by adding a term in Γ_7 and a gauge interaction. By using the method of partial exact solution outlined above an equation with an su(2)-symmetric pseudoscalar interaction is obtained again in $M^{3,1}$.

These, and further extensions of (3.2,7) will be studied in more detail in a forthcoming paper. Here we wish only to show how C_6S.G., like C_4S.G., is an appropriate geometrical frame where the equations of the most familiar physical phenomena may be easily accommodated and, consequently, some of their still unexplained features, like internal symmetry, may find geometrical explanation.

4. C_{2n} SPINOR GEOMETRY. QUANTUM THEORY.

In the original geometrical defintion the 2^n spinor components ξ_α are numbers; they may represent, in a classical theory, spinorial charges. In a quantized theory they should be substituted by operators which, because of statistics should, in general, obey some anti-commutation relations. Let us try to determine them.

Reversing equations β) and γ) we may express the monomials $\xi_\alpha \xi_\beta$ in terms of linear combinations of tensor components $v^{(n)}_{[a_1 \ldots a_n]}$ and $V^{(p)}_{[a_1 \ldots a_p]}$. They build up the symmetric matrix $\xi\xi^T$. In the quantized theory, because of statistics, the diagonal elements $\xi^2_\alpha (\alpha=1,2 \ldots 2^n)$ will have to map every state vector to zero. We may then assume in the quantized theory the stronger ansatz:

$$\xi^2_\alpha = 0 \quad , \quad \alpha = 1,2,\ldots,2^n \, . \tag{4.1}$$

We may wish to generalize this equation with the SO(2n)-covariant (for n=2 it will mean relativistic, for n = 3 conformal) equations for the quantized charges $\varphi^{I,II}_\alpha$:

$$[\varphi^I_\alpha, \varphi^I_\beta]_+ = 0 \quad ; \quad [\varphi^{II}_\alpha, \varphi^{II}_\beta]_+ = 0 \quad , \quad \alpha,\beta = 1,2,\ldots,2^{n-1}, \tag{4.2}$$

from which (4.1) follows as a particular case. We will then assume the hypothesis:

Hypothesis 4,1

In a quantized theory spinor components belonging to the same semi-spinor anticommute.

If we adopt this hypothesis we see from (2,1β) that tensor $V^{(n)}$ in β) are zero when expressed in terms of quantized ξ_α, while $V^{(p)}$ in γ) may be non-zero and may represent Bose tensor operators in terms of spinorial ones. We have then a distinction between essentially classical bilinear spinor polinomial β) and both classical and quantum mechanical ones γ). This fact stresses the already mentioned substantial difference between the tensors $V^{(n)}$ and $V^{(p)}$ defined in β) and γ) respectively.

With hypothesis 4,1 C_{2n} S.G. acquire a simple and significant form in the quantized theory:

4.1 Quantized C_4 S.G.

Reversing (3.1,1) we obtain in the quantized theory (eq. (3.1,1β) gives no contribution because of hypothesis 4.1):

$$[\psi_\alpha, \psi_\beta]_+ = (\gamma_2 \gamma_0 \gamma_5 \gamma_\mu)_{\alpha\beta} V^\mu ,$$

which is, formally, the characteristic equation of Poincaré superalgebra (provided ψ_α are Majorana spinors). However we have here only the geometrical frame of Poincaré superalgebra since we cannot interpret the operators V^μ as generators of Poincaré translations since C_4 spinor geometry is only that one of the light cone.

It is interesting to note that $f_{\mu\nu} = \tilde{f}_{\mu\nu} = 0$ in the quantized theory which means that we cannot interpret the fields $\bar{E}$, $\bar{H}$ in terms of quantized spinors: this would violate Hypothesis 4,1 and consequently statistics, and it is known in fact that statistics is the main difficulty for the neutrino theory of light.

4.2 Quantized C_6S.G.

Reversing the equations (3.2,6) (eq. (3.2,5) gives no contribution because of hypothesis 4,1) giving $[\varphi_I]\cap[\varphi_{II}]$, we obtain, in terms of the two Dirac spinor $\psi^{(1)}$ and $\psi^{(2)}$ components contained in ξ:

$$[\psi_\alpha^{(2)}, \psi_\beta^{(j)}]_+ = (\gamma_0 \gamma_2 \gamma_5 \gamma_\mu)_{\alpha\beta} P^{(j)\mu} \delta_{ij}, \; i,j = 1,2, \quad (4.2,1)$$

that is formally very near to the characteristic equation of Poincaré extended supersymmetry.

But, from (3.2,6) we may also obtain the full conformal superalgebra equation [5]

$$[\xi_\alpha, \xi_\beta]_+ = \frac{1}{8}\{B[\Gamma_a,\Gamma_b]\}_{\alpha\beta} J^{ab} + a B_{\alpha\beta},$$

where a = 0 when ξ is defined as a pure spinor in C_7.

5. CONCLUSION

We see that if we take pure semi-spinors as the elementary geometrical objects we derive the following results:

I) Spinors of C_{2n} Clifford algebras with n > 3 (and probably n ⩾ 4) are needed in order to build up the tensor geometry of Minkowski space-time; which means that we need spinor multiplets with more than 2 (and probably 4) Dirac spinors to build up space-time.

II) Already the simple C_4S.G. and C_6S.G. , even if not allowing the construction of the whole tensor geometry of space-time contain quite naturally the geometrical frame for the representation of the most elementary physical phenomena like light propagating in vacuum and the propagation of massless quanta.

III) The equations (2.5) and (2.7) which establish the equivalence between pure spinors and n-dimensional isotropic planes of E_{2n} may be interpreted as wave equations and it is interesting that already in C_4S.G. they give the geometrical frame for both Weyl and Maxwell equations, while in C_6S.G. they may be partially, but exactly, solved and give rise to equations presenting formally an su(2) symmetric interaction.

It is easy to predict that C_{2n}S.G. with $n \geqslant 4$ (necessary to generate $M^{3,1}$ tensor geometry) will give us equations with su(N) internal symmetry with $N = 2(n-2)$, reduced by the pureness constraint for the spinors: these become important for $n > 3$. This reduction will, in general, generate <u>families</u> of su(2) internal symmetries (since the high dimensional spinor will be reducible to a multiplet of C_6 spinors).

IV) Supersymmetry equations (extended) arise from C_{2n}S.G. by the equations expressing bilinear spinor monomials in terms of linear polinomials in the "new" E_{2n} tensors $[\varphi_I] \cap [\varphi_{II}]$ generated by C_{2n}S.G. provided hypothesis 4.1 is satisfied.

We feel that hypothesis 4.1, if it resists further critical study (and will be possibly derived from some basic axioms), may be deeply rooted in the geometrical basis of mechanics since it determines the border between purely geometrical and quantum mechanical elements and their co-existence.

REFERENCES

[1] E. Cartan, The Theory of Spinors (Herman, Paris 1966).

[2] J. Rzewuski, Bull Acad. Polon. Sci. 6, 261 (1958);
R. Penrose and M.A.H. MacCallum, Physics Reports 6C (4) (1973);
R. Ablamowicz, J. Mozrzymas, Z. Oziewicz, J. Rzewuski, Reports on Mathematical Physics 14, 89 (1978).
R. Penrose, R.S. Ward, Einstein Centennial Volume (1979).

[3] Y. Murai, Nucl. Phys. 6, 489 (1958);
R.L. Ingraham, "Free Field Equations of Conformal Relativity in Riemannian Formalism. II. - Some Explicity Solutions", New Mexico State University preprint, and refs. therein (1981).

[4] P. Furlan, ISAS Preprint, in preparation.

[5] M. Daniel and C.N. Ktorides, Nucl. Phys. B115 (1976).

ELEMENTARY PARTICLES IN UNIVERSAL SPACE-TIME*

Interim Report

Irving E. Segal
Massachusetts Institute of Technology
Cambridge, Mass., USA

and

Institut für Theoretische Physik
TU Clausthal
Federal Republic of Germany

* The author thanks the Humboldt foundation for facilitating this work, and the National Science Foundation for partial support.

Abstract: Elementary particle theory is studied in a natural, essentially unique and maximal space-time $\mathbf{M}$ related to Minkowski space by a 4-dimensional relativistic version of stereographic projection. The natural energy operator in $\mathbf{M}$ exceeds that in $\mathbf{M}_0$ by unobservably little in highly localized states, but by a physically quite significant amount in sufficiently delocalized states. The difference in energies drives a theoretical redshift in excellent agreement with systematic galaxy and quasar observations. Here the possible implications for microscopic physics are treated on the basis of general considerations of causality, symmetry, and stability. Fields and particles in $\mathbf{M}$ are closely related to counterparts in $\mathbf{M}_0$, facilitating unique development of theory based on the analysis of bundles on $\mathbf{M}$ derived from conventional ones on $\mathbf{M}_0$. The classification of states is however subject to greater unification in $\mathbf{M}$. The mass in $\mathbf{M}$ differs from the mass in $\mathbf{M}_0$ for delocalized states, which fact could reconcile the difference in observed mass with the fundamental identity of the muon and electron. Among further differences, parity does not commute with a central element of the fundamental symmetry group of $\mathbf{M}$ in a manner indicative of parity non-conservation.

Introduction

The success of quantum and microscopic particle theory has left fundamental physics with a split personality. On an intuitional basis, both micro- and macro- theory have aspects of compelling simplicity and observational validity. This is true to such an extent that each theory can be represented as in principle _the_ fundamental one. At a logical level however each theory _has_ basic problems going back to its origins that remain unresolved. Moreover in both regimes the experimental or observational work increasingly outstrips the predictive power of the traditional theories, unless they are modified by ancillary observationally-indicated hypotheses that create additional basic problems.

There is a technique for dealing with this situation that has been emphasized by Minkowski and Dirac, and is exemplifield by major works of Einstein, Heisenberg, and Schrödinger, but has nevertheless not had widespread application in contemporary physical investigations. This is, briefly, the reconstruction of a _verifiable_ theory from nature on the basis of modern _mathematical ideas_. I mean not merely the use of cleaner and more precise mathematical technology, highly desirable and useful as that is, _or_ the refinement of generalities remote from observational confrontation, however interesting this may be from mathematical or philosophical standpoints.

Rather I mean the application of whatever mathematics appears naturally appropriate to the concrete and quantitative comprehension of the real objective physical world. Such mathematics may be elementary or sophisticated, classical or largely undeveloped; the issue is to coordinate it with fundamental physical theory, refining, manufacturing, or developing it as need be. Some historians ascribe the great progress of the Kepler, Newton, Lagrange, Maxwell, etc. eras to just such an enterprise, and correlate the progress with the availability both of sophisticated mathematics _and_ new experimental information. Thus in practice as well as in theory, this approach to the problems of fundamental physics may be just as powerful as the continuing construction of larger and more specialized accelerators, telescopes, and the like.

My aim here is not to explore history, for which I have no special competence, but to describe a design for the mathematical equivalent of one such observational apparatus. Objectively speaking it has been quite successful in cosmology, although it has not yet been widely adopted by theorists,- for no scientific reason. If as seems difficult to doubt the universe is a unified entity, its basic laws being applicable over a continuous spectrum of distance scales from the ultramicro- to ultramacroscopic, the same apparatus should be useful in predicting elementary particle phenomena. Thus I plan to indicate how a synthesis of very general conceptual physics with natural and coherent mathematical correlatives may lead to a more intelligible theory of greater predictive power.

Universal Physics

By this I mean physics formulated in such a way that it applies universally to any of a general class of space-times. But what is space? time? space-time?

My formulation will ultimately merge basically with that of the great macroscopists (notably Mach and Einstein), but their work was prior to the advent of particle theory, whose standpoint I take initially. Subject to a postiori validation, I begin by simply postulating a given cosmos, by which I mean a 4-dimensional manifold M endowed with the most general properties of causality and symmetry. (Cf. refs. 1,2) M is interpreted physically as an 'empty' or 'reference' space-time structure. Later (below) this 'bare' cosmos will become 'clothed' by the interactions of its energetic contents, which will endow it additionally with a particular metric structure. Mathematical theory tells us that M must be locally Minkowskian (to admit appropriate causal symmetries implementing isotropy), and that there exists a unique universal cosmos **M** in which all others are contained,- Minkowski space $\mathbf{M_0}$, de Sitter and anti de Sitter spaces, etc.

Specifically, **M** is the homogeneous space **SU**(2,2)/**P**, where boldface letters denote the universal covers, i.e. the simply connected objects that are locally identical to those designated by the corresponding non boldface letters, P denoting the scale-extended (11-dimensional) Poincaré group. There is an essentially unique notion of causality in **M** that is invariant under the action of the group **G** = **SU**(2,2); and the familiar special cosmos mentioned are simply the orbits in **M** under subgroups of **G**; these subgroups are precisely the causal symmetry groups of these subcosmos, as **G** essentially is for **M**.

If we want to do physics on the cosmos M, which can be regarded as a subcosmos of **M**, we need to specify what is 'time', 'energy', 'particle', 'quantum number', 'interaction', 'S-matrix', to begin with. Other necessary concepts are largely corollary to such basic ones, from which predictions as to energy levels, S-matrix elements, and other in principle observable aspects can be derived. To make the empirical connection, these concepts must be related to the quantities 'observed' in laboratories. The quotation marks here refer to the need to allow properly for the model-dpendence, to a greater or lesser extent, of all observation. In particular, in the procedures for data reduction, the use of widely prevalent models tends to become increasingly unconsious. For example, the 'spin' of a particle is 'measured' by experiments that are planned and analyzed with 'particle' modelled as a cross-section of a Minkowski-space bundle of a particular type. When analyzed assuming other cosmos and/or bundles, different results are to be expected, or different experiments may be called for. This is not at all to say the model is arbitrary. The issue of the (or 'a') correct model is not necessarily a relative one, as in the classic question of which of two objects is in motion, since a postiori goodness-of-fit tests can non-trivially accept or reject a truly scientific model. In **M** for example there should be 'conformal' effects, above and beyond 'relativistic' ones, that are observable with sufficient precision, just as the fine structure of hydrogen is a 'relativistic' effect, absent 'non-relativistically'.

The most basic of the concepts enumerated above have been treated elsewhere, and I can here only describe some key aspects. If one wants to reconcile time as a coordinate with time as a group parameter (typical usages in macro- and microphysics), the time concept is highly restricted, but is never-the less not entirely unique locally. All the cosmos enumerated above are locally Minkowskian, thus causally locally identical, and hence admit the same local 'covariant' clocks, where this is a clock measuring time restricted in the foregoing sense. But it is not the case that any two local covariant clocks are equivalent, within conjugacy by a local causality-preserving transformation. It becomes a matter of physics to determine which clock applies in a given observational situation.

For simplicity I now confine the discussion to the only two cosmos that admit global splittings into time and space: $\mathbf{M}_0$ and $\mathbf{M}$. The only global clock on $\mathbf{M}_0$ is the usual one, for which the quantum-mechanical energy in its universal form is $-i\partial/\partial x_0$. (In any particular system, this operator takes on a specific form.) On **M** the only global clock is non-conjugate locally to that on the imbedded $\mathbf{M}_0$, and its energy operator (i.e. the generator of infinitesimal temporal evolution, apart from the factor $-i$) has the form $H = H_0 + H_1$, where $H_0 = -i\partial/\partial x_0$ and H_1 is the transform of H_0 under conformal inversion. As a vector field in $\mathbf{M}_0$, H takes the form (in natural units, in which $\hbar$, c, and an ultramacroscopic fundamental unit of length, describable as the 'radius of the universe', are 1)

$$H = -i(1-\tfrac{1}{4}x^2)\partial/\partial x_0 + i(x_0/2)(x_0\,\partial/\partial x_0 + x_1\,\partial/\partial x_1 + x_2\,\partial/\partial x_2 + x_3\,\partial/\partial x_3) \quad ; \; x^2 = x_0^2 - x_1^2 - x_2^2 - x_3^2.$$

One estimates from this that for a photon of frequency n, the expected values of H_0 and H_1 are respectively n and $N^2/4n$, where N is the number of oscillations in space for the photon wave function. The latter quantity can become quite large for a photon that has been propagated over a cosmic distance, although initially, i.e. for a localized photon, H_1 will have an unobservably small value. Laboratory measurements can determine only n, and not N, and thus are measurements of H_0.

The chronometric redshift theory is based on the idea that the true driving hamiltonian is nevertheless the larger energy H. If a photon state is initially $\emptyset$, at time t its wave function $\emptyset_t$ will be $e^{itH}\emptyset$, and its observed frequency will be the expected value of H_0 in the state $\emptyset_t$, i.e. $(H_0\emptyset_t, \emptyset_t)$, a non-constant function of t; and not $(H\emptyset_t, \emptyset_t)$, which is constant as a function of t but is not directly observable.

A rigorous analysis based on the Maxwell equations and the associated unitary representation of the conformal group shows that $(H_0\emptyset_t, \emptyset_t) = (1+z)^{-1}(H_0\emptyset, \emptyset)$, where $z = \tan^2(t/2)$, apart from terms that are unobservably small, for frequencies n and redshifts z that are even remotely in the observable range. A striking feature of this theoretical result is that the resulting redshift z depends only on t, and not at all on the frequency n; this is at the same time one of the most remarkable properties of the observed redshift. From the foregoing prediction of the redshift z one derives by elementary calculus predicted non-parametric relations between observed quantities (as t above is not) that are found to be in excellent agreement with the observations reported in all objectively specified galaxy or quasar samples that are amenable to a reasonably conservative statistical study. Such studies show also that the observations are inconsistent with the predictions of the expanding-universe hypothesis, irrespective of the values given its adjustable parameters, unless extremely rapid and rather complex 'evolution' is postulated. Such evolution reduces the predictive power of the expanding-universe hypothesis virtually to nil in the case of quasars, and in the case of galaxies, the Hubble law fails.

H_0 is of course scale-covariant, in the sense that under the change of scale that carries x into ax, H_0 transforms into aH_0. On the other hand, H_1 is anti-scale covariant, transforming into $a^{-1}H_0$; and H does not transform simply at all under scale transformations. The notions of scale- and anti-scale covariance are naturally relative to a particular time and place of observation. The cosmic redshift thus appears as a flow of energy betwen two observationally-defined forms, that inevitably accompanies the delocalization of its wave function as required by Maxwell's equations.

Let us call H_0 the flat and H_1 the anti-flat energy. The commutation relations between H, H_0, H_1, and the other generators of the causal group of **M** (i.e. the group of all causality-preserving transformations on **M**) are uniquely determined by the specification of the three fundamental units. With the specification made above, all constants of structure are $\pm$ 1 or 0. Minkowski's 'program' (cf. ref. 3) can be regarded as completed by the theoretical model just described that is based on the action of **G** on **M**. The basic physical units are determined in an intrinsic group-theoretic manner, so that the constants of structure have the simple values stated. In addition **G** is terminal from the standpoint of Minkowski's desideratum of the replacement of a given physical symmetry group by a simpler one: the Lie algebra of **G** (i.e. **G** as an infinitesimal group) can not be a limiting case (or 'deformation' or 'contraction') of any inequivalent group of the same number of parameters.

Micro-physics

Given any putative cosmos, theoretical microphysics on it can be set up on the basis of quite general considerations that may be outlined as follows.

1) Let M be the cosmos, assumed a 4-dimensional manifold endowed with a causal structure. Given M, its fundamental symmetry group G, consisting of all causal transformations on M, is determined. Unless G is fairly large, there is no apparent natural and experimentally practical way to label states; in the cases earlier cited, G fulfils this requirement, implementing isotropy and homogeneity, and providing a satisfactory set of state labels. If G is fairly large, there are two subgroups that play a special role: (i) the isotropy group G_0 fixing a generic point (in practice, taken as the point of observation); (ii) the 'mass-conserving' subgroup G_1. G_0 is

uniquely determined mathematically, but G_1 can only be suggested by mathematical considerations and must ultimately be determined by physical ones. The mathematical considerations include: existence of a physical vacuum state for associated quantized fields (in general, non-compact semi-simple groups can not be expected to leave such a state of interacting fields invariant); existence of a universal energy operator generating temporal evolution, and having a semi-bounded spectrum for suitable systems (this last is a fairly stringent constraint, ruling out for example such a group as O(4,1) as G_1, since no generator of this group has a semi-bounded spectrum in any infinite-dimensional representation); existence of G_1-invariant 'wave equations', serving to propagate causally in accordance with the causal structure given on M (or deduced from a given conformal Lorentz structure), initial data given on a space-like surface in M (i.e., essentially 'finite propagation velocity'), and to decompose into subspaces with such G_1-invariant and finite propagation velocity characteristics, a G-invariant bundle representing a species of massive particle.

As examples, for $\mathbf{M}_0$, G is the scale-extended Poincaré group, of dimension 11; G_0 is the scale extended (homogeneous) Lorentz group; G_1 is the usual 10-dimensional Poincaré subgroup. Note that G_1 admits a unique invariant Lorentz metric, relevant especially in macrophysics.

For **M**, G is the universal cover of the conformal group, locally O(2,4), modulo a two-element subgroup of its center that acts trivially on **M**, strictly speaking. (However, we systematically take space-time manifolds in their simply-connected forms, and also their symmetry groups, as suggested by theoretical physical experience, and as serves only to enlarge slightly, although on occasion quite significantly, the correlative physical system.) G_0 is the scale-extended Poincaré group (more exactly, its universal, two-fold, cover, when G is taken simply connected). From a mathematical standpoint, G_1 may logically be chosen as any of O(2,3), the Poincaré group, or the maximal compact subgroup K of G. All of these groups admit unique invariant Lorentz metrics on **M**. However K is mathematically the most natural, being the only one that is transitive on **M** (in its simply connected form, in which it is 'essentially' but not strictly compact), and the only one for which the existence of stationary states is assured; and there are cogent physical indications for K (cf. below). However, O(1,4) could be used by giving up positivity of the energy in its direct form and adopting physical interpretations specific for this group, as discussed by Philips and Wigner, and by Schrödinger, among others. The group O(2,3) admits positive-energy actions in the general sense, and has been used in cosmological theory by a group in Peking associated with the work of L. K. Hua.

2) Next one must determine or select the bundles that will model physical fields and particles. In conventional relativistic theory these are induced from finite-dimensional representations of G_0. Since the G_0 for $\mathbf{M}_0$ is a factor group of the G_0 for **M**, every bundle of the conventional type on $\mathbf{M}_0$ gives rise to a corresponding bundle on **M**. The latter bundle, when restricted to $\mathbf{M}_0$ as imbedded in **M**, is identical to the original bundle. Thus there is a very simple direct relationship between conventional relativistic fields and corresponding fields on **M**; but these corresponding fields, or bundles mathematically speaking, exhibit phenomena that do not occur in the conventional bundles, and would be theoretically physically significant, within the framework of the choiceof **M** as underlying cosmos.

3) Having fixed the bundle(s), one must determine the stable, by which I mean 'positive energy', invariant subspaces under G. Physically stable particles must be modelled by such bundle subspaces. This imposition of a 'spectral condition' is readily effected in $\mathbf{M}_0$ by Fourier analysis, which however is inapplicable in other cosmos. Mathematical investigation reveals however that their bundles do in interesting cases possess stable subspaces appropriate for modelling physically indicated fields.

4) Once a suitable mass-conserving subgroup has been determined or chosen, the stable subspaces will decompose on restriction to it into further subspaces, each defined by a mass parameter and a wave operator for the stable subspace as a whole. For example, the analog on **M** of the Klein-Gordon equation is briefly discussed below, and in refs. 4 and 5.

5) For detailed physics in the direction of empirical predictions it is invaluable and probably essential to have explicit decompositions of the stable subspaces, and/or their fixed-mass subspaces, defined by physically appropriate complete sets of quantum numbers. In the case of **M**, for example, there is a natural set of quantum numbers that approximates the conventional relativistic ones, in the 'flat' local limit in which **M** approximates $\mathbf{M}_0$, which is informally describable as 'the limit when the radius of the universe becomes infinite'. Cf. refs. 4 and 5 regarding these quantum numbers.

6) Causal interactions, whether G-invariant or only G_1-invariant, are determined by suitable invariants of the inducing representations of the bundles involved (or equivalently, corresponding tensor product decompositions). The procedure indicated by Brauer and Weyl in ref. 6 in the case of $\mathbf{M}_0$, quite succinctly, can be correlated with theoretical physical practice, and extended to non-trivial bundles such as those on **M**.

7) Quantization remains to be effected. Invariant methods applicable to linear aspects of general groups and bundles have been available for some decades in the more mathematical literature. Because of the incomplete reducibility of the bundles on **M** relatively sophisticated techniques are required, even after restriction to the stable subspaces, such as C*-algebra formulations; with suitable additional structure derived either from a specific physical interpretation or mathematical restrictions, these are reducible to more familiar types of linear vector space formats, involving specific representations of the canonical commutation or anti-commutation relations.

Non-linear G- or G_1- invariant interactions must be expected to appear as divergent as corresponding ones in relativistic field theory. The first-order perturbation theory will be essentially finite and useful; but the interaction hamiltonian or lagrangian as a whole will show ultraviolet divergences analogous to those in the original Dirac radiation theory. On the other hand the problem of infra-red divergences will disappear in the case of **M** at least. In this cosmos there are other significant simplifications: the infinite volume limit is not necessary, space being compact; exp (-sH) is typically of finite trace, without approximation; the problem of the limit in particle scattering theory as the time becomes infinite is eliminated, since infinite Minkowski times correspond to finite times in **M.**

8) A technical problem that is important from a practical point of view is that of convenient parallelization of the bundles involved. Moreover the treatments of the foregoing aspects must be adapted to this parallelization. Otherwise computations become mathematically quite burdensome and physically abstract, and the explicit relationship to the flat limit, which is essential for the physical interpretation since the flat quantities are those measured in elementary particle experiments, is difficult to express.

9) In any event, in connection with the last observation or otherwise, suitable approximate agreement with the 'relativistic' conventional limit must be verified.

What is involved in 1) - 9) is fairly elaborate, being rather more complex than in the case of conventional relativistic theory, regarded as an extension of non-relativistic theory. It nevertheless appears technically accessible. To attempt to convey a more concrete impression of procedures, I describe aspects of the treatment of conventional fields of given spin on **M**, and especially the scalar case.

In describing these fields on **M** there is the complication that the spin spaces at different points are really different, and have no unique natural identification, as they do in the case of a linear space such as $\mathbf{M}_0$. However, there is an identification of different spin spaces that is mathematically convenient and physically appropriate; namely the K-invariant parallelization. To effect parallelization, one can take any point p_0 of **M**, thought of as the point of observation, and any simply transitive group F of causal transformations acting in a neighborhood of p_0. One then identifies the spin space at an arbitrary point p with that at p_0 by using the transformation in F that carries p into p_0.

Thus the usual 'flat',- unconscious,- parallelization describes **M** locally as $\mathbf{M}_0$, and uses the group of vector translations on $\mathbf{M}_0$. This does not work globally on **M**; the flat translation group is transitive only on $\mathbf{M}_0$, and not at all on **M**. However, it describes laboratory observations on particles as presently conducted, data reduction being in flat (Poincaré invariant) terms and experimentation confined to a quite localized region.

For a global parallelization on **M** it is convenient to use the group of left translations on **M**, regarding **M** as the group **U**(2); right translation would not be essentially different. RESULT: The total infinitesimal action of **G** = **SU**(2,2) on **M** is the sum of its external (geometrical) action, i.e. a vector field, with an internal (spin) part, i.e. a matrix (which here may depend on the space-time point) that can be given explicitly.

Specifically, the spin component of the action of the element X of the Lie algebra $\underline{G}$ of **G** on the fields (i.e. bundle cross sections), transforming according to the infinitesimal representation r of the isotropy group $\mathbf{G}_0$ is r(Y), where Y is the following element of $\underline{\mathbf{G}}_0$:

$$Y = \tfrac{1}{4}\,\mathrm{tr}\,(a+bZ^{-1}-cZ-d) + \begin{pmatrix} -Z^{-1}b+cZ+d & -Z^{-1}b \\ -cZ & d \end{pmatrix}$$

where X is represented by the matrix $\begin{pmatrix} a & b \\ c & d \end{pmatrix}$ in su(2,2),

identified with the Lie algebra of **G**, and Z is arbitrary in U(2), with which **M** is locally identified. At Z = I, one has the conventional spin component, but only at this point, in general.

This is the action as it looks on a neighborhood of the point of observation,- a large neighborhood, comparable in size to $\mathbf{M}_0$,- and it extends simply and canonically to all of **M**. It should be noted that the decomposition into internal and external parts of the action of a group element X is quite materially parallelization-dependent, and without invariant physical significance. Thus in the indicated parallelization vector translations in $\mathbf{M}_0$ have non-vanishing spin components, which moreover depend on the space-time point. But the parallelization greatly facilitates practical computations, making it possible to treat the fields as ordinary vector-valued functions.

As an example, consider the scalar bundle on **M**. Here the inducing representation R is identically 1 on the Poincaré subgroup of G_0, and on the scale transformation that carries the vector x into ax takes the form $R(a) = a^n$, for some fixed real 'weight' n. It is a theorem that there is a unique value of n (n=1) for which the corresponding bundle has a non-trivial positive-energy subspace; and for which this subspace has a non-trivial **G**-invariant subspace.

To describe the subspace explicitly, one needs to observe that each of the three putative mass-conserving subgroups of **G** earlier indicated has a corresponding mass operator. This is the usual d'Alembertian in the case of the Poincaré subgroup, and closely related to the quadratic Casimir operator in the case of the other subgroups, which are simple (Lie-theoretically). The three mass operators all differ in the large, but are identical infinitesimally, i.e. at the observer's position. Locally, as regards particle aspectsthey in general differ by little, for states that are sufficiently localized the difference being unobservably small. The bundle is however locally rigorously independent of the choice of $\mathbf{G}_1$; and the bundle subspace annihilated by each of the mass operators is the same, and in fact **G**-invariant.

The situation is similar in the case of higher spins, and particularly in the physically important case of spin 1/2. In both the spin 0 and spin 1/2 case the invariant subspace is uncomplemented by a **G**-invariant subspace; there are however complements relative to each of the three putative mass conserving subgroups, which differ from subgroup to subgroup. In the spin 1/2 case the invariant 'zero-mass' subspace is naturally identified physically with the space of neutrino wave functions on **M**. It is interesting that on **M** there is a clear distinction between a neutrino and an electron of zero mass, both mathematically and in terms of the natural physical interpretation. The electron wave function is in the quotient of the stable spin 1/2 subspace, modulo the neutrino subspace. The lowest mass in both subspaces is physically effectively zero, although strictly non-vanishing, of the order of the space curvature. The physical electron mass that is non-vanishing is to be regarded in accordance with Mach's principle as not an individual attribute of the electron, but rather its interaction energy with the rest of the energetic contents of the universe. The quotient space is mathematically much larger in a certain sense than the invariant neutrino subspace, and is naturally interpreted physically as the space of all electrons, including the muon , tauon, and so on. These higher electrons thus appear as having the same K-invariant mass as an ordinary electron, but a different P-invariant mass, as observed, as is possible if their wave functions are sufficiently extended. Thus one arrives at a substantially parameter-free description of higher electrons as species of long-tailed ordinary electrons, a concept in qualitative agreement with the more parametric proposal of Dirac, from which the hadron bag model evolved; see ref. 7.

A significant difference between the different spins on **M** is their invariance features under a new universal symmetry that appears. The transformation ρ that acts on space as S^3 by sending each point into its antipode, and on time t by sending it into $t + \pi$, is in the center of the group **G**. Being in the center, it would be expected to be a constant, an absolute invariant, for any given type of field. It acts differently in the spin 0,1/2, and 1 cases, acting entirely trivially on Maxwell photon wave functions, and least trivially in the spin 1/2 case. The central element ρ has in addition the interesting property of not being invariant under parity, but of being transformed into a different element of the center. For photons, both ρ and its transform act trivially, and there is no parity anomaly, but in the spin 1/2 case their actions differ in sign. This is suggestive of maximal parity non-conservation, and indeed a parity doublet formulation providing the simplest parity-invariant formulation of a 4-fermion coupling on **M** has as its local flat limit the chiral-invariant maximal parity-violating interaction in $\mathbf{M}_0$ in its original form.

At this point we could return to macrophysics and interpret Mach's principle by correlating H_1 with gravitational energy, and the Equivalence Principle as **G**-invariance, but space limitations prevent us from more than noting that the K-invariant formulation of **M** is naturally identifiable with the Einstein Universe; see ref. 8.

References

1. I. E. Segal (1976). Mathematical cosmology and extragalactic astronomy. Academic Press, New York.
2. -------- (1980). Time, energy, relativity, and cosmology. In Symmetries in science, ed. B. Gruber and R. S. Millman, Plenum Publ. Corp, New York, pp. 385-396.
3. H. Minkowski (1908). Raum und Zeit. Adress at 80th Assembly of German Natural Scientists, Cologne, 21 Setp. 1908.
4. I. E. Segal, H. P. Jakobsen, B. Ørsted, S. M. Paneitz, and B. Speh (1981). Covariant chronogeometry and extreme distances: Elementary particles. Proc. Natl. Acad. Sci. 78, 5261-5265.
5. S. M. Paneitz and I. E. Segal (1982). Analysis in space-time bundles. To appear in Jour. Funct. Analysis.
6. R. Brauer and H. Weyl (1935). Spinors in n dimensions. Amer. Jour. Math.
7. P. Hasenfratz and J. Kuti (1978). Phys. Rep. 40C, No.2, 75.
8. C. Møller (1971). The theory of relativity (2nd ed.), Clarendon Press, Oxford, Eng.

SECOND ORDER TANGENT STRUCTURE FOR SPACETIMES

C.T.J. Dodson

Department of Mathematics
University of Lancaster
England

M.S. Radivoiovici

Department of Mathematics
University of Bucharest
Romania

ABSTRACT: It is fundamental to the concept of a smooth manifold M that it has the canonical tangent vector bundle TM. This, geometrically, is a bundle of directions represented by curves equivalent up to velocity and analytically it is a bundle of derivations represented by partial differential operators. Related constructions can be made, for second order operators and for curves equivalent up to acceleration, but there the analytic and geometric approaches do not give equivalent structures. Various bundles of second order on a spacetime are compared and some natural structures are noted, including an acceleration-sensitive semimetric.

1. INTRODUCTION

Given a smooth manifold M, the tangent vector bundle over M is also a smooth manifold TM, and hence so also is TTM over TM, and so on. Geometrically we view TM as the space of all possible velocities, or tangent vectors, for curves in M but analytically it is more convenient to handle as the space of all first order derivations on real functions defined on M. Formally, we interpret T as a functor from the category of smooth manifolds to the category of vector bundles (cf. Ref.(3) for example):

$$T \;:\; \mathrm{Man} \rightarrow \mathrm{Vbun} \;:\; f\Big\downarrow\begin{matrix}M\\ \\ M'\end{matrix} \;\;\mapsto\;\; \begin{matrix}TM\\ \Big\downarrow\\ TM'\end{matrix}\; Df$$

Here Df (or f_* in some texts) is the derivative of f, which is at each point a linear approximation to f.

There are then two ways to generalise TM, each yielding a second order structure over M. Geometrically we obtain T^2M, the space of all possible accelerations for curves in M, and analytically we obtain J^2M, the space of all second order derivations on functions on M. It turns out that J^2 is always a functor from Man to Vbun (cf. Ref.(1)) but T^2 is only such a functor in the presence of a linear connection on M (cf. Ref.(4)). Moreover, when there is a linear connection on M then the canonical vector bundle TTM over TM can be decomposed as a vector bundle over M (cf. Ref.(7)).

For a spacetime (M,g) the Lorentz structure g determines the unique symmetric compatible connection ∇^g in the frame bundle LM and in its reduce subbundles. Hence both T^2M and J^2M are available as vector bundles over a spacetime. As is to be expected, the extra structure of a connection on M makes T^2M more simple with fibre $\mathbb{R}^8$ than J^2M which has fibre $\mathbb{R}^{14}$. These vector bundles are associated respectively to principal bundles L^2M (cf. Ref. (4)) and P^2M (cf. Ref. (14)), both admitting reduction via g.

We shall compare various bundles of second order. They have significance for a spacetime because of the importance attached in physics to differential equations of second order. Geometrically, a differential equation of first order on M is a section of TM , that is a tangent vector field on M . Correspondingly, a differential equation of second order on M is a section of TTM , that is a tangent vector field on TM . The presence of a linear connection simplifies the geometry by reducing the number of essential dimensions.

In the following we suppose that (M,g) is a spacetime. So M is a connected, Hausdorff, non-compact, paracompact smooth manifold with Lorentz structure g with respect to which M is time and space oriented and maximally extended[10].

2. CONSTRUCTION OF SECOND ORDER TANGENT BUNDLES

We consider a spacetime (M,g) of dimension n .

At any $x \in M$ the <u>k-ject space</u> J^k_xM to M at x is the vector subspace of real linear maps given by

$$J^k_xM = \{v \in L(F_x;\mathbb{R}) : F^c_x \cup (F^o_x)^{k+1} \leqslant \ker v\}$$

where

$$F_x = \{C^\infty f : N_x \to \mathbb{R} : N_x \text{ some open set about } x\}$$

$$F^c_x = \{f \in F_x : f \text{ is constant on some open set about } x\}$$

$$F^o_x = \{f \in F_x : f(x) = 0\}$$

$$(F^o_x)^{k+1} = \{\text{finite sums of products of } (k+1) \text{ elements from } F^o_x\} .$$

Then J^1_xM is an isomorph of T_xM , the usual tangent space to M at

x. We obtain vector bundles over M from

$$_{x\varepsilon M} J^1_x M = J^1 M \cong TM = {}_{x\varepsilon M} T_x M$$

$$_{x\varepsilon M} J^k_x M = J^k M \quad \text{, the \underline{k-jet bundle} over } M \ .$$

Of particular interest for us are $J^1M \cong TM$ with fibre $\mathbb{R}^n$, and J^2M with fibre $\mathbb{R}^{\frac{1}{2}n(n+3)}$.

Next consider at any $x\varepsilon M$ the curves

$$C_x = \{C^\infty f:(-\varepsilon,\varepsilon) \to M : f(o) \quad x, \quad \text{some real } \varepsilon > 0\} \ .$$

Velocity and acceleration fields along these are given respectively by

$$\dot{f} \ : \ (-\varepsilon,\varepsilon) \to TM \ : \ t \longmapsto D_t f \ (1)$$

$$\ddot{f} \ : \ (-\varepsilon,\varepsilon) \to TTM \ : \ t \longmapsto D_t \dot{f} \ (1)$$

These immediately suggest a classification for the curves in C_x via equivalence relations $\sim_x$ and $\approx_x$ where

$$f \sim_x h \iff \dot{f}(o) = \dot{h}(o)$$

$$f \approx_x h \iff \ddot{f}(o) = \ddot{h}(o) \quad \text{(and then also } \dot{f}(o) = \dot{h}(o)\text{)} \ .$$

So we obtain the quotients $C_x/\sim_x$ and $C_x/\approx_x$ which give vector spaces $T^1_xM = \mathbb{R}^n$ and $T^2_xM = \mathbb{R}^{2n}$. These in turn determine the <u>velocity bundle</u> T^1M and the <u>acceleration bundle</u> T^2M . In fact, the velocity bundle is isomorphic to the tangent bundle TM and, since (M,g) has a natural linear connection ∇^g , we also have T^2M isomorphic to $TM \oplus TM$ by (4). Dombrowski[7] showed that a linear connection on M reduces the bundle TTM to a vector bundle of structure $TM \oplus TM \oplus TM$ over M.

Given a chart (U,ϕ) about $x\varepsilon M$ then we obtain the following frames from derivations with respect to coordinates:

$(\partial_i)_y$ for T_yM , $y\varepsilon U$

$(\partial_i \oplus \partial_j)_y$ for T^2_yM , $y\varepsilon U$

$(\partial_i \oplus \partial_k \oplus \partial_k)_y$ for TT_yM , $y\varepsilon U$.

3. PRINCIPAL BUNDLES

Denote by $G^1(n)$ the general linear group $Gl(n;\mathbb{R})$. Then:

$$TM \cong (LM\times\mathbb{R}^n)/G^1(n)$$

shows that TM is an associated bundle to the principal bundle LM , hence so are J^1M and T^1M . Similarly, we can interpret T^2M as an associated bundle

$$T^2M \cong (L^2M\times\mathbb{R}^{2n})/G^1(2n) \quad . \qquad \text{(Cf. Ref.(4))}$$

Note that although $T^2M \cong TM \oplus TM$, the structure group of L^2M is $Gl(2n;\mathbb{R})$

For each $k=1,2,...$ it is possible (cf. Ref.(14)) to interpret J^kM as an associated vector bundle:

$$J^kM \cong (P^kM\times\mathbb{R}^{n_k})/G^k(n) \quad .$$

Here, in particular,

$$P^1M = LM \ , \quad n_1 = n \quad \text{and} \quad n_2 = \tfrac{1}{2}n(n+3) \quad ,$$

and $G^k(n)$ consists of classes of diffeomorphisms, of a neighbourhood of the origin in $\mathbb{R}^n$, equivalent up to k-th order derivatives (cf. Refs.(5),(11) and (14) for local expressions).

In our case of a spacetime (M,g) with Levi Civita connection ∇^g , some consequences for the second order structures are as follows

i) $T^2M \cong TM \oplus TM$. (Cf. Ref.(4))

ii) ∇^g determines a unique section of $P^2M/G^1(n)$. (Cf. Ref.(13))

iii) ∇^g determines a $G^1(n)$ subbundle of P^2M . (Cf. Ref.(12))

iv) ∇^g in LM determines a unique connection $\tilde{\nabla}^g$ in L^2M . (Cf. Ref.(4))

v) L^2M (like LM) is parallelisable. (Cf. Ref.(5))

vi) $\tilde{\nabla}^g$ determines a covariant derivation on T^2M . (Cf. Ref.(4))

vii) $\tilde{\nabla}^g$ is compatible with $\tilde{g}$, the life of g to a scalar product on T^2M , and $\tilde{\nabla}^g$ coincides with $\nabla^{\tilde{g}}$ as a derivation. (Cf. (Cf. Ref.(4))

viii) $\tilde{\nabla}^g$ determines unique second order horizontal lifts of vector fields and curves. (Cf. Ref.(4))

Hennig[12] showed that a conformal structure and a symmetric linear connection on M determine a Weyl structure ("parallel transport preserves light cones") if and only if their corresponding subbundles of P^2M have non-trivial intersection. He deduced that given a conformal structure and a projective structure that are compatible, then the choice of a distinguished class of autoparallel paths is sufficient to select a unique symmetric linear connection from the projective class such that it yields a Weyl structure. Hence one obtains the "physical" Lorentz structure up to a constant factor.

Schmidt[17] used ∇^g to define a Riemannian structure

$$g' = \theta\cdot\theta + w^g\cdot w^g$$

on O^+M, the $SO^+(1,3)$ subbundle of LM, where θ is the basic 1-form and w^g is the connection form. From this he constructed a topological completion, the <u>b-completion</u> $\bar{M} = M \cup \partial M$, where ∂M is the <u>b-boundary</u> for spacetime (M,g); detailed calculations using this construction can be found in Ref.(2). Corresponding forms $\tilde{\theta}$ and $\tilde{w}^g$ arise on L^2M, giving a Riemannian structure g'' on the second order frame bundle. However, the essential structure of L^2M is carried by a $G^1(n) \oplus G^1(n)$ subbundle and it was pointed out by C.J.S. Clarke that the natural quotient completion so found coincides with the b-completion.

4. ACCELERATION STRUCTURE AND SINGULARITIES

It was shown[4] that T^2M is a vector bundle over TM with fibre $\mathbb{R}^n$, that is the bundle of covariant accelerations over the manifold of velocities. This allows a decomposition:

$$T^2M \longrightarrow TM \oplus TM \ : \ [f]_{\approx x} \longmapsto (x, \dot{f}(o) \oplus \nabla^g_{\dot{f}(o)}\dot{f}) \ .$$

Now, given a timelike curve c in (M,g), then $\nabla^g_{\dot{c}}\dot{c}$ is spacelike and measures the obstruction to c developing as a geodesic; in physical terms for a particle, this is the obstruction to being free. Suppose that c is parametrized by proper time, that is:

$$c : [o,\tau_m) \longrightarrow M \quad \text{and} \quad \forall\tau\epsilon[o,\tau_m) \quad g(\dot{c}(\tau),\dot{c}(\tau)) = -1$$

We call such a curve <u>τ-incomplete</u> if it is inextensible and τ_m is finite. A τ-incomplete curve is called <u>b-incomplete</u> if any (and hence every one) of its horizontal lifts is of finite length in (O^+M,g'); b-incomplete curves have endpoints on the b-boundary ∂M.

Rosso[15] showed that it is sufficient for a τ-incomplete curve to have bounded Lorentz-norm acceleration:

$$\left\|\overset{g}{\dot{c}}\dot{c}\right\| = \left|g(\dot{c},\dot{c})\right|^{\frac{1}{2}}$$

in order to be b-incomplete. On the other hand[16], it is impossible to have b-incomplete curves with unbounded $\|\nabla^{g}_{\dot{c}}c\|$. However, when c is τ-incomplete, then it will be b-incomplete if

$$\int_{o}^{\tau_m} \exp \int_{o}^{\tau} \|\nabla^{g}_{\dot{c}}\dot{c}\|$$ is finite. (Cf. Ref.(6))

In the case of a geodesic c, this integral reduces to τ_m, the proper time.

The notion of b-incompleteness can be extended to any curve in (M,g) that is inextensible with finite horizontal life; indeed the b-boundary provides endpoints for all such curves, not just the timelike ones. Suppose then that c is any inextensible curve in M and

$$c_e : [o,\beta_e) \longrightarrow M$$

is this curve, parametrized by the arc length of c_e, the unique horizontal life of c to $(0^+M,g')$ through some frame e over $c(o)$. The group $SO^+(1,3)$ determines all possible such parametrizations. For it is isomorphic to the fibre of 0^+M over $c(o)$, , which determines all possible initial orthonormal frames through which to effect the lift. Now, $SO^+(1,3)$ is known to act uniformly continuously on $(0^+M,g')$ so given orthonormal frames e,e' at $c(o)$ then there is some $\gamma\varepsilon SO^+(1,3)$ such that:

i) $R_\gamma(e') = e$ (by transitivity of the right action on fibres)

ii) $\beta_e = r_\gamma \beta_{e'}$ for some real $r_\gamma \geqslant o$.

The <u>b-length</u> of c is the infimum over possible lifts:

$$b(c) = \inf_{\gamma\varepsilon SO^+(1,3)} \{\beta_e \ : \ e = R_\gamma(e') \text{ for some } e'\varepsilon\, 0^+_{c(o)}M\}$$

where with components $X_i(t)$ of $c(t)$ with respect to $e(t)$, the parallel propagation of e from $c(o)$ to $c(t)$,

$$\beta_e = \int_{\mathrm{dom}(c)} \left(\sum_0^3 (x_i)^2\right)^{\frac{1}{2}}$$

These properties suggest generalising our earlier integral to arbitrary curves by defining the a-length:

$$a(c) = \inf_{\gamma \varepsilon SO^+(1,3)} \left\{ \int_o^{\beta_e} \exp \int_o^{\beta} \| \nabla^g_{\dot{c}_e} \dot{c}_e \| \right\} .$$

From the properties of the exponential function and the non-negativity of $\| \nabla^g_{\dot{c}_e} \dot{c}_e \|$ it follows that $a(c) \geqslant b(c)$. Equality holds for curves with null covariant acceleration, in particular for geodesics. However, $a(c) = o$ if $b(c) = o$ and it is well known that $b(c)$ can be zero for non-constant curves, for example in closed Friedmann spacetime[2]. So, like the b-length, the a-length determines only a semimetric on M itself:

$$d_a \; : \; M \times M \longrightarrow [o,\infty)$$

$$: \; (x,y) \longmapsto \inf \{a(c) : c \text{ is from } x \text{ to } y\} .$$

It follows that the unwanted identification of physically distinct singularities in the b-boundary would not be avoided by using a-lengths instead of b-lengths. On the other hand, the a-length may be useful in better separating those events joined by curves of high acceleration. The difference between a-length and b-length provides an invariant measure of the intrinsic effort required by a particle (or a rocket) in following such a trajectory, in the timelike case.

ACKNOWLEDGEMENTS

The authors would like to thank Professor Abdus Salam, the International Atomic Energy Agency, and the United Nations Educational Scientific and Cultural Organisation for hospitality at the International Centre for Theoretical Physics, Trieste, Italy, during their visits there in the summer of 1981.

REFERENCES

1. Ambrose, W., Palais, R.S. and Singer, I.M. "Sprays", Anais da Academeia Brasileira de Ciencias 32, 2 (1960) 163-178.
2. Dodson, C.T.J., "Space-time Edge Geometry" Int.J.Theor.Phys. 17, 6 (1978) 389-504.
3. Dodson, C.T.J., Cateogries, Bundles and Spacetime Topology, Shiva, Orpington, 1980.
4. Dodson, C.T.J. and Radivoiovici, M.S. "Tangent and frame bundles of order two", Analele Stiintifice ale Universitatii "Al. I. Cuza" din Iasi, Sect. Matematica (in press).
5. Dodson, C.T.J. and Raivoiovici, M.S. "Second-order tangent structures", Int.J.Theor.Phys. (in press).
6. Dodson, C.T.J., Sulley, L.J. and Williams, P.M. "An acceleration condition for b-incomplete timelike curves", Math.Proc. Cambridge Philos. Soc., 90 (1981) 191-193.
7. Dombrowski, P. "On the geometry of the tangent bundle", J. Reine und Ang. Math., 210 (1962) 73-88.
8. Ehresmann, C. "Les prolongements d'une variété différentiable", (Parts I and II) C.R.Acad.Sci.Paris (1951) 598-600 and 777-779.
9. Ehresmann, C. "Introduction à la théorie des structure infinitésimals et des pseudo-groupes de Lie", Colloque de Géométrie Differentielle de Strasbourgh, Centre National de la Recherche Scientifique Paris, (1953) 97-110.
10. Hawking, S.W. and Ellis, G.F.R., The Large Scales Structure of Space-Time C.U.P. Cambridge, 1973.
11. Hennig, J.D. "G-structures and spacetime geometry" Part I: Geometric objects of higher order," Preprint IC/78/46 (1978) I.C.T.P., Trieste Italy, and Part II: "Conformal and projective structure imply Weyl structure," Preprint, Inst.Theor.Phys.Univ.Clausthal (1978).
12. Hennig, J.D. "Jet bundles and Weyl geometry," Lecture Notes in Physics 139 edited by H.D. Doebner, Springer-Verlag, Berlin 1981.
13. Kobayashi, S. "Canonical forms on frame bundles of higher order contact", Proc.Symp.Pure Math. Vol. 3, American Math.Soc. (1961) 186-193.
14. Kobayashi, S. Transformation groups in differential geometry, Springer-Verlag, Berlin 1972.
15. Rosso, F. "A comparison between Geroch and Schmidt space-time incompleteness", Rend.Accad.Sci.Fis.Mat.Napoli , Ser.IV,43 (1976) 362-371.
16. Rosso, F. "Bundle and proper incompleteness with unbounded world acceleration in general relativity ", Boll.Un.Mat.Italiana A5,16 (1979) 593-597.
17. Schmidt, B.G. "A new definition of singular points in general relativity", Gen.Relativ.Gravit. 1, (1971) 269-280.

FIRST COHOMOLOGY GROUPS FOR LIE GROUPS

S.R. Komy
Department of Mathematics
Riyadh University, Saudi-Arabia

First Cohomology groups for Lie groups and Lie Algebras , with values in a representation space ,are introduced .If the group is simply connected ,then the first Cohomology groups for the Lie group and its Lie algebra are isomorphic.If the group is connected but not simply connected ,then the first Cohomology group is sensitive to the representation of its centre.

In the theory of Lie groups, both the infinitesmal and the global points of views present themselves alternately.In many cases ,the global properties are more or less reflected in the infinitesmal ones which makes the Lie algebra an excellent tool to study properties in the large .However there are some global properties of which it is not known how they are related to the infinitesmal properties .One of these relationships which has not been fully investigated is the connection between the Lie algebra Cohomology[1)],and the group Cohomology[2)]with values in a given representation space .

Of particular interst , are the first Cohomology groups of the Poincare' group relative to its various actions on sets of solutions of free relativistic wave equations .Each first order Cocycle wil correspond to a finite energy solution carring an interinsic charge[3)].Using any Cocycle found,a fully relativistic quasi-free Quantum field theory can be constructed which will describe many attributes of particles .

In the following ,let G be a connected Lie group ,E be its Lie algebra , φ be a continuous representation in the vector space V,and let θ ($\equiv d\varphi$) be the induced representation of E in V. A 1-Cocycle for G is a smooth map $f:G \longrightarrow V$,defined by:

$$f(xy) = \varphi(x)\, f(y) + f(x) \quad ;x,y \in G .$$

The set $Z^1(G,V)$ of 1-Cocycles form an additive abelian group .

Each vector $a \in V$,defines a 1-Cocycle called a 1-Coboundary , defined by:

$$f_a(x) = \varphi(x)\, a - a \quad ; x \in G .$$

Let $B^1(G,V)$ denote the set of 1-Coboundaries .The first Cohomol -ogy group $H^1(G,V)$ is defined as the quotient group $Z^1(G,V) /B^1(G,V)$.

Similarly, a 1-Cocycle for E is a linear map $F: E \longrightarrow V$, defined by :

$$F([h,k]) = \theta(h)\, F(k) - \theta(k)\, F(h) \quad ; h,k \in E .$$

The set $Z^1(E,V)$ of 1-Cocycles form an additive abelian group . Also ,each vector $a \in V$ defines a 1-Cocycle called a 1-Coboundary, given by :

$$F_a(h) = \theta(h)\, a \quad ; h \in E .$$

If the set of 1-Coboundaries is denoted by $B^1(E,V)$, then the first Cohomology group $H^1(E,V)$ for E is the quotient group $Z^1(E,V)/B^1(E,V)$.

If the group G is compact ,then $H^1(G,V) = 0$[4) .But if G is not compact ,then this result cannot be implemented .This is due to the non-existence of an invariant Haar measure .If G is a simply connected group ,then the computation of its first Cohomology group is reduced to an algebraic one , and we have the following theorem :

THEOREM 1:

Suppose that G is a simply connected group ,and E is the corresponding Lie algebra ,then $H^1(G,V)$ is isomorphic to $H^1(E,V)$. The proof of this theorem is a consequnce of the following lemma and propositions (for proofs see [5]) :

Lemma :

Suppose that G is a connected group ,and let $f \in Z^1(G,V)$ such that $f'(e) = 0$. Then $f = 0$.

Proposition 1 :

Let $f \in Z^1(G,V)$,and define the linear map $F: E \longrightarrow V$, by : $F(h) = f'(e;h)$,$h \in E$. Then $F \in Z^1(E,V)$.

Proposition 2:

Suppose that G is a simply connected group ,and let $F \in Z^1(E,V)$. Then there exists an $f \in Z^1(G,V)$ such that $f'(e) = F$.

If G is connected but not simply connected ,we consider its covering group $\tilde{G}$ .Then G is of the form $\tilde{G}/N$,where N is a normal discrete subgroup(the centre) .If φ is a representation of $\tilde{G}$, then φ is also a reoresentation of G .Then we have the following

THEOREM 2:

If G is a connected group with centre N ,G is the universal covering group of G ,and φ is a continuous representation in a vector space V , then the two cases follow:

case 1: If N is not represented trivially , then $H^1(G,V) = 0$.

case 2: If N is represented trivially ,then $H^1(G,V) \cong H^1(\tilde{G},V)$.

The proof of this theorem is a direct consequence of the Cocycle relation .

The Poincaré group is a semi-direct product of a solvable subgroup(the translations) ,and a semi-simple subgroup (the Lorentz group) . Araki[6)] reduced the problem of computing the first Cohomology group of any connected Lie group to that of its semi-simple subgroup . A special case is the Poincaré group in 2 space-time dimensions ,which is a solvable group and Araki's reduction is trivial .Computations of the first Cohomology groups for the Poincaré group in this special case as well as in 3 and 4 space-time dimensions will be presented else where .

Refrences :

1- G.Hochschild and J.P.Serre ,Ann. Math. ,vol. 57,no.3, 1953 .

2- Eilenberg and MacLane ,Ann.Math. ,vol. 44 ,no. 1, 1947 .

3-S.R.Komy ,Proceedings of the 15 th. annual conference of Computer sciences and Mathematics ,Cairo Univ. ,December 1980 .

4-K.Parthasarathy and K.Schmidt ,Lecture notes in Math. ,vol.272.

5-S.R.Komy ,Proceedings of the 1980 Conference on differential Geometric methods ,Clausthal (to appear) .

6-H.Araki ,Publ.RIMS,Kyoto Univ.,vol.5,1969/1970 ,361-422 .

IV. QUANTUM FIELD THEORIES AND GAUGE THEORIES

ALGEBRAIC GAUGE QUANTUM FIELD THEORY ON GENERALIZED KALUZA-KLEIN SPACES

H.D. Doebner and F.B. Pasemann
Institut für Theoretische Physik
TU Clausthal
Federal Republic of Germany

Abstract: A quantum field theory for gauge fields reflecting the topological and geometrical structures of classical gauge theories is given in terms of complete sets of n-point functions on generalized Kaluza-Klein spaces. The quantization procedure leads to linearized theories for non-abelian gauge groups. In the abelian case, ie. G=U(1), the resulting theory recovers standard results.

1. Introduction

There are beautiful topological and geometrical structures inherent in classical gauge theories which become most apparent in the differential geometric formulation [1] - [5] of these theories. Quantum field theories (QFT) for gauge fields in general do not reflect these structures in a direct way.

The quantization procedure outlined here is an attempt to transport the topological and geometrical structures of the classical theories to the quantum level, thus giving explicit conditions for the n-point functions of the resulting gauge QFT. It makes use of an algebraic approach to axiomatic QFT's which is close to the Borchers formulation [6] in the scalar field case. A complete set of n-point functions is represented by a functional C over an involutive, topological algebra B constructed as tensor algebra over an appropriate test form space. Field strengst functionals C^{Ω} and potential functionals C^{ω} live on test form algebras B^{Ω} and B^{ω} which are by construction distinct.

The geometrical properties of the classical objects, as well as the information contained in their classical equations, is reflected in the definition of corresponding subalgebras or ideals in the test form algebras. The topology of the classical theory will determine the properties of maps relating different subalgebras and ideals, respectively.

From this algebraic version of a gauge QFT an operator formalism can be constructed by using a slightly generalized GNS-construction [7]. For the potentials an indefinite metric formalism is then obtained.

The quantized theory is formulated in a global and gauge covariant way on the generalized Kaluza-Klein space [5], [8] of the corresponding classical theory.

In the following the quantization procedure is outlined and it is shown how it applies to pure gauge theories. Finally there is a short discussion of the main aspects of this algebraic gauge QFT.

2. Classical gauge theories

The differential geometric formulation [1] - [5] is shortly rewiewed. We fix a compact, connected, semi-simple Lie group G with Lie algebra $\mathbb{G}$, and we will consider only pure gauge field theories over Minkowski space. Such a G-gauge theory can be represented by a pair (P,ω) where $P(M,G)$ is a principal G-bundle over Minkowski space (M,g), g the Minkowski metric, and where ω denotes a connection form on $P(M,G)$ representing the classical gauge potential.

We provide the total space P of $P(M,G)$ with a pseudo-Riemannian metric $\hat{g}$ defined by

$$\hat{g}(X,Y) := \pi^{*}g(X,Y) + \tilde{g}(\omega(X),\omega(Y)) \quad , \qquad X,Y \in \mathcal{X}(P),$$

where $\tilde{g}$ is the unique biinvariant positive definite metric on G given by the negative of the Cartan-Killing form on G, and $\pi : P \longrightarrow M$ is the bundle projection.

The space $(P,\hat{g})$ is called a generalized Kaluza-Klein space.

Let $A^p_G(P,\mathbb{G})$ denote the space of $\mathbb{G}$-valued p-forms on P, G-equivariant with respect to the adjoint representation. The following operators act on $A^p_G(P,\mathbb{G})$: the exterior derivative d, the horizontal projection H^* [9] with respect to ω, the Hodge star operator $*$ [10] with respect to $\hat{g}$, the co-derivative δ defined by $\delta := (-1)^p *^{-1} d *$, where $*^{-1}$ is given here by $*^{-1} := (-1)^{p(n-p)+1} *$, the covariant derivative $\nabla := H^* d$, and the covariant coderivative $\not\nabla := H^*\delta$.

In terms of these operators the classical equations are given in the form

(1) $$\Omega := \nabla\omega ,$$

(2) $$\nabla\Omega = 0 ,$$

(3) $$\not\nabla\Omega = 0 ,$$

where $\omega \in A^1_G(P,\mathbb{G})$ is the connection form, $\Omega \in A^2_G(P,\mathbb{G})$ is the corresponding curvature form representing the classical field strength, equation (2) is the Bianchi identity, and equation (3) is the physical field equation for a pure gauge field theory. In addition ω and Ω have the following properties:

(4) $$H^*\omega = 0 ,$$

(5) $$H^*\Omega = \Omega ,$$

where (4) shows, that ω is a vertical 1-form, whereas (5) expresses the fact that Ω is a basic 2-form. (Recall that $\alpha \in A^p_G(P,\mathbb{G})$ is called basic, iff $i(Z_h)\alpha = 0$, $h \in \mathbb{G}$, where Z_h denotes the fundamental vector field on P generated by $h \in \mathbb{G}$.) The space of basic p-forms on P is denoted by $A^p_B(P,\mathbb{G})$, and $H^* : A^p_G(P,\mathbb{G}) \longrightarrow A^p_B(P,\mathbb{G})$ is a projection.

Equations (1) - (3) are equivalent to the usual equations for the corresponding local (gauge dependent) A and F fields on Minkowski space (M,g) [11], and properties (4) and (5) are reflected by the gauge transformation properties of these fields.

3. Quantization

The construction of an algebraic QFT for gauge fields originates from the observation that an axiomatic QFT is completely determined by its system of n-point functions, i.e. vacuum expectation values of products of field operators.

The Borchers approach [6] to the scalar field case considers the n-point functions to be continuous linear functionals on an involutive topological algebra B constructed as a tensor algebra over the space V of test functions, i.e.

$$B := \sum_{r=0}^{\infty} \oplus (\otimes^r V) \quad .$$

For specific functionals on B (corresponding to the Wightman functions) the algebraic theory is then equivalent (via the reconstruction theorem) to the standard operator (Wightman-) theory.

In the following we fix a generalized Kaluza-Klein space $(P,\hat{g})$. The "quantization rule" leading to an associated gauge QFT is given by two steps:

Step 1: The classical fields we are considering are represented by elements $K \in A^p_G(P,\mathbb{G})$. Motivated by the fact that the map $* : A^{n-p}_G(P,\mathbb{G}) \longrightarrow A^p_G(P,\mathbb{G})$ is an isomorphism, $n = \dim P$, we choose as basic test form spaces the spaces $D^p_G(P,\mathbb{G})$ of $\mathbb{G}$-valued, G-equivariant p-forms on P with compact support. The associated test form algebras B^p are constructed from these spaces as involutive topological algebras

$$B^p := \sum_{r=0}^{\infty} \oplus (\otimes^r D^p_G(P,\mathbb{G})) \quad .$$

A complete set of n-point functions is represented by a continuous linear functional $C^K \in (B^p)'$, $K \in A^p_G(P,\mathbb{G})$, where $(B^p)'$ denotes the dual space of B^p.

An operator $\varkappa : D^p_G(P,\mathbb{G}) \longrightarrow D^q_G(P,\mathbb{G})$, $\varkappa = d, H^*, \ldots$ induces a homomorphism $\theta_\varkappa : B^p \longrightarrow B^q$ given by

$$\theta := \sum_{\tau=0}^{\infty} \oplus \ (\otimes^\tau \varkappa) \quad ,$$

and its transpose map $\theta'_{\varkappa'} : (B^q)' \longrightarrow (B^p)'$, where $\varkappa'$ is the metric adjoint operator of $\varkappa$ (e.g. $\varkappa = d$, $\varkappa' = \delta$).

Closed subalgebras $B^q_\varkappa \subset B^q$ are defined by $B^q_\varkappa := \mathrm{Im}\, \theta_\varkappa$, and closed two sided ideals $I^q_\varkappa \subset B^q$ are generated by $\varkappa$-exact forms $\alpha \in D^q_G(P,\mathbb{G})$, i.e. α is of the form $\alpha = \varkappa\beta$, $\beta \in D^p_G(P,\mathbb{G})$.

Step 2: To transfer the information contained in the classical equations to the algebraic level, we distinguish between two cases:

a.) The classical equation gives a relation between fields $K \in A^p_G(P,\mathbb{G})$ and $R \in A^q_G(P,\mathbb{G})$, i.e.

$$\varkappa K = R \quad . \tag{6}$$

We then postulate that the equivalent relation holds for corresponding field functionals $C^K \in (B^p)'$ and $C^R \in (B^q)'$:

$$\theta'_\varkappa C^K = C^R \quad .$$

But since

$$\theta'_\varkappa C^K(f) = C^K(\theta_{\varkappa'} f) \quad , \quad f \in B^q \ ,$$

equation (6) distinguishes the subalgebra $B^p_{\varkappa'} \subset B^p$.

b.) The classical equation is of the form

$$\varkappa K = 0 \qquad , \qquad K \in A_G^p(P,\mathbb{G}) \ .$$

In this case we postulate that a field functional $C^K \in (B^p)'$ vanishes on the ideal $I^p_{\varkappa'} \subset B^p$, where the operator $\varkappa'$ is the metric adjoint of $\varkappa$.

Example: As an example we will apply this quantization rule to the pure gauge field theory described in chapter 2: The classical field strength is represented by a curvature form $\Omega \in A_G^2(P,\mathbb{G})$. As basic test form space for a quantum field strength description we therefore choose the space $D_G^2(P,\mathbb{G})$, and construct from this the test form algebra B^2. The n-point functions of the (field strength) theory will be represented by functionals $C^{\Omega} \in (B^2)'$, the basic properties of which are determined as follows.

i.) Quantization of equation (5), i.e. horizontality of Ω, is given by the postulate

$$\theta'_{H^*} C^{\Omega} = C^{\Omega} \ ,$$

which shows that C^{Ω} "lives" on the subalgebra $B^2_{H^*} \subset B^2$, i.e. it vanishes on ker θ_{H^*} .

ii.) Before quantizing the field equation (3) we shall realize that in general the field equation reads

$$\nabla \Omega = J \qquad , \qquad J \in A_B^1(P,\mathbb{G}) \ ,$$

where J is a current 1-form. A pure gauge theory is defined by J = 0. Quantization of (3) then gives

$$\theta'_{\nabla} C^{\Omega} = 0 \ ,$$

i.e. C^{Ω} vanishes on the subalgebra $B^2_{dH^*} \subset B^2$.

iii.) Quantization of the Bianchi identity (2) will establish the property that C^{Ω} vanishes on the ideal $I^2_{\delta H*} \subset B^2$.

To summarize we give the following

Definition: A field strength functional C^{Ω} is a functional $C^{\Omega} \in (B^2)'$ satisfying

a.) $\quad \theta'_{H*} C^{\Omega} = C^{\Omega}$,

b.) $\quad C^{\Omega}$ vanishes on $B^2_{dH*} \subset B^2$,

c.) $\quad C^{\Omega}$ vanishes on $I^2_{\delta H*} \subset B^2$.

Applying the quantization rule to the classical potential theory we obtain the test form algebra B^1 constructed from the basic test form space $D^1_G(P,\mathbb{G})$, and we end up with the

Definition: A potential functional C^{ω} is a functional $C^{\omega} \in (B^1)'$ satisfying

a.) $\quad C^{\omega}$ vanishes on $I^1_{H*} \subset B^1$, $\quad$ (from equation (4)),

b.) $\quad C^{\omega}$ vanishes on $I^1_{\delta\nabla H*} \subset B^1$, $\quad$ (from equation (2)),

c.) $\quad C^{\omega}$ vanishes on $B^1_{\delta\nabla H*} \subset B^1$, $\quad$ (from equation (3)).

Quantization of the defining equation (1) for the curvature form Ω will give the relation between the field strength functionals C^{Ω} and corresponding potential functionals of the quantized theory, i.e.

$$\theta'_{\nabla} C^{\omega} = C^{\Omega} .$$

This relation leads to the subalgebra $B^1_{\delta H*} \subset B^1$ which is called the distinguished potential algebra. (It will play a crucial role in the definition of the physical state space for the corresponding operator theory.)

To define an appropriate gauge QFT, the functionals $C^{\omega} \in (B^1)'$ and $C^{\Omega} \in (B^2)'$ have to satisfy some additional properties like lokality, spectral condition, etc. This will be discussed elsewhere.

4. Final Remarks

For the construction of the algebraic theory we fixed a generalized Kaluza-Klein space $(P,\hat{g})$, i.e. we fixed a connection form ω on $P(M,G)$. The discussion of the ω-dependence of the formalism involves the discussion of gauge degrees of freedom, and this will be given in a subsequent paper.

Since M was chosen to be the Minkowski space, the principal G-bundle $P(M,G)$ is trivializable. Interesting topological aspects will show up in the theory, if non-trivial principal G-bundles (e.g. for monopol theories) are considered.

Although the theory was formulated on $(P,\hat{g})$, it equivalently may be given in terms of objects on the base space M (here Minkowski space), but in the non-abelian case its structure is much more complicated there.

That the theory is linearized in the non-abelian case may be observed from the fact, that the classical interaction terms $[\![\omega,\omega]\!]$ and $[\![\omega,\Omega]\!]$ are absorbed into the operator ∇, i.e. no use was made of the structure equations [9]

$$\nabla\omega = d\omega + \tfrac{1}{2}[\![\omega,\omega]\!] \quad , \quad \nabla\Omega = d\Omega + [\![\omega,\Omega]\!] \quad .$$

The constructed theory therefore describes a quantum gauge field interacting with its classical counterpart.

Using a slightly generalized GNS-construction as described in [7] a corresponding operator theory may be generated from the algebraic theory. Two state spaces, $\mathcal{H}^{\Omega}$ and $\mathcal{H}^{\omega}$, will be obtained: $\mathcal{H}^{\Omega}$ for the field strength decription, $\mathcal{H}^{\omega}$ for the potential description. $\mathcal{H}^{\omega}$ in general carries an indefinite metric, but on a physical subspace $\mathcal{H}^{\omega}_{ph} \subset \mathcal{H}^{\omega}$, which is essentially determined by the distinguished potential algebra $B^{1}_{\delta H*}$, it will be positive semi-definite. The spaces $\mathcal{H}^{\omega}$, $\mathcal{H}^{\omega}_{ph}$ and $\mathcal{H}^{\omega}_{o}$ (the subspace of null-vectors in $\mathcal{H}_{ph}$) and their properties resemble the situation characteristic for the Gupta-Bleuler formalisme of the free Maxwell theory.

References

[1] Trautman, A. : Rep.Math.Phys. 1 (1970) 29

[2] Daniel, W., Viallet, C.M. : Rev.Mod.Phys. 52(1980)175

[3] Eguchi, T., Gilkey, P.B., Hanson A.J.: Phys.Rep. 66 (1980) 214

[4] Atiyah, M.F. :"Geometry of Yang-Mills Fields", Pisa 1979

[5] Hermann, R. : "Yang-Mills, Kaluza-Klein, and the Einstein Program", MathSciPress, Brookline 1978

[6] Borchers, H.J. : Algebraic Aspects of Wightman Field Theory, in Sen, R., Weil, C. (eds.): "Statistical Mechanics and Field Theory", Halsted Press, New York 1972

[7] Bongaarts, P.J. : Maxwell's equations in Axiomatic Quantum Field Theory, pt. I: J.Math.Phys. 18(1977)1510, pt II: J.Math.Phys. (to appear)

[8] Kerner, R. : Ann.Inst.H.Poincaré 9(1968)143 , and J.Math.Phys. 21,(1980)2553

[9] Greub, W., Halperin, S., Vanstone, R. : "Connections, Curvature, and Cohomology", Academic Press, New York 1973

[10] Flanders, H. :"Differential Forms", Academic Press, New York, 1963

[11] Popov, D.A. : Theor.Math.Phys.(engl.transl.) 24(1976)879

ULTRA-VIOLET CANCELLATIONS IN SUPERSYMMETRIC GAUGE QUANTUM FIELD THEORIES

S. Ferrara

CERN, Geneva, Switzerland

ABSTRACT: We consider some ultra-violet suppressions of quantum divergences occurring in global supersymmetry and supergravity. The possible relevance of these cancellations in the context of unified theories is briefly discussed.

The ultimate goal of high energy physics is to have a consistent unified picture of all fundamental interactions. Supersymmetric gauge theories offer at present a possible scenario to describe the low energy world as well as the physics at the Planck scale where gravitational interactions become strong.

The so-called hierarchy problem [1] of grand unified theories of electroweak and strong interactions can in principle have a natural explanation [2] in the context of N = 1 (one supersymmetry generator) supersymmetric Yang-Mills theories because of the existence of some non-renormalization theorems [3]. In particular these theorems forbid the usual Higgs doublets, responsible for the breaking of the Weinberg-Salam $SU(2)_L \times \times U(1)_Y$ gauge group, to suffer from quadratic mass renormalization, thus preventing the theory from an unnatural explanation of a low energy scale $M_W \gtrsim 100$ GeV not mixed with a huge (unification?) scale $M_X \gtrsim 10^{15}$ GeV or M_{Planck} through radiative corrections.

In the context of local (gauge) supersymmetry, supergravity theories [4] have the chance of describing a consistent (finite?) theory of gravity and eventually they offer the ultimate scheme for a superunification of gravitational interactions with the non-gravitational strong and electroweak forces. In particular the hidden local SU(N) symmetry discovered in N extended supergravities suggests [5] that our low energy physics be described in terms of states belonging to composite supermultiplets of the elementary preonic fields of the basic supergravity Lagrangians [6].

The main interest raised by supersymmetric gauge theories, apart from their intrinsic elegance and their profound algebraic structure, is due to their exceptional ultra-violet properties. The ultra-violet behaviour is improved with respect to conventional renormalizable theories because of mutual cancellations of boson and fermion loops, due to restrictions on coupling and masses imposed by supersymmetry. The absence of quadratic mass renormalization for scalar partners of chiral fermions in super-Yukawa theories has been proven since the birth of supersymmetry [3]. The general proof for the absence of quadratic mass renormalization in the physically interesting situation of gauge theories, which is directly related to the hierarchy problem of GUTs, was obtained soon after [7].

Using superspace techniques [7,8] one can show that the usual Weinberg power counting formula for the superficial degree of divergence of ordinary Feynman diagrams is inadequate if applied to supersymmetry theories. When power counting is established in superspace the cancellations between boson and fermion loops are automatically taken into account and the lowering of independent renormalization constants is easily understood. Unfortunately, this technique is not yet available in N extended supersymmetry where more spectacular cancellations seem to occur. In particular it is generally believed that the maximally extended supersymmetric theories reveal additional convergence properties due to the PCT self-conjugate nature of their elementary supermultiplets. More specifically, it has been shown by explicit calculations [9] that in the $N = 4$ Yang-Mills theory there is no infinite charge renormalization [$\beta(g)=0$] in the first three-loops. Recently, arguments based on the $N = 1$ properties of the $N = 4$ theories have been used to show the vanishing of the β function of the $N = 4$ Yang-Mills theories to all orders of perturbation theories. One argument [10] is based on the relation of the trace anomaly to the axial anomaly and then to the vanishing of the latter because of the absence of a U(1) chiral symmetry in $N = 4$ Yang-Mills theory due to its SU(4) global symmetry. A more convincing argument, which uses only the preservation of four supersymmetries to higher loops, uses the non-renormalizability theorem of intrinsically chiral interactions (superpotential term) of $N = 1$ Yang-Mills theories and the fact that $N = 4$ supersymmetry relates (actually identifies) the gauge coupling with a superpotential coupling [11].

The same argument has been given [12] to prove the vanishing of the β function of $N = 4$ conformal supergravity where the group SU(4) is gauged. The vanishing of the β function in $N = 4$ conformal supergravity is known to occur by explicit calculation [13].

This argument also applied to higher N extended supergravities. The main point is to regard these theories as $N = 1$ supersymmetric theories and to use the $N > 1$ supersymmetries to relate different $N = 1$ invariant terms which have different renormalization properties in $N = 1$ supersymmetry. This trick has been used very recently [12] to give arguments for the vanishing of the SO(N) β function to all orders in SO(N) gauged supergravities with $N > 4$. It is not impossible that similar arguments can be used in SU(8) extended supergravity. It is worth mentioning that the failure of these criteria by explicit calculation would imply that $N > 1$ supersymmetric theories are plagued by anomalies and that the only quantum mechanical supersymmetry algebra would be $N = 1$ supersymmetry.

The vanishing of the one-loop $\beta(g)$ function in $N > 4$ extended supergravities [14] with a gauged SO(N) group and the absence of the cosmological constant renormalization Λ, due to the supersymmetry relation $\Lambda \sim g^2/K^4$, can also be understood from purely kinematic sum rules [15] related to the helicity spectrum of extended particle supermultiplets

$$\sum_{\lambda}(-)^{2\lambda}d(\lambda)\lambda^{p} = 0 \qquad p < N \tag{1}$$

$$\sum_{\lambda}(-)^{2\lambda}d(\lambda)\lambda^{p}C_{2}(\lambda) = 0 \qquad p + 2 < N \tag{2}$$

When $d(\lambda)$ is the multiplicity of the state and $C_2(\lambda)$ is the quadratic SO(N) [or SU(N)] Casimir of the corresponding SO(N) [or SU(N)] representation. Incidentally, Eq. (1) also explains the vanishing of the one-loop β function in N = 4 Yang-Mills theory because the one-loop β function for a particle of helicity λ can be parametrized as follows

$$\beta_{\lambda} \sim (-)^{2\lambda}C_{2}(\lambda)(a + b\lambda^{2}) \tag{3}$$

where a,b are λ independent coefficients and $C_2(\lambda)$ is λ independent in the N = 4 Yang-Mills theory.

Another ultra-violet cancellation occurring in extended supergravity, is connected to the mass formulae of spontaneously broken supergravity through dimensional reduction [6]. These mass relations imply that the one-loop cosmological term is finite in $N \geq 6$ supergravities and is crucial to prove that these theories are one-loop renormalizable.

In an N extended supergravity, spontaneously broken through dimensional reduction from five dimensions, one has [17]

$$\sum_{J}(-)^{2J}(2J + 1)\mathcal{M}_{J}^{2k}(m_i) = 0 \qquad 2k < N \tag{4}$$

where $\mathcal{M}_J$ is the mass of the particle of spin-J of the graviton supermultiplet and m_i are the parameters (i=1,...,4 for N=8) introduced through dimensional reduction which set up the scale of the supersymmetry breaking.

It is remarkable that the spin and mass sum rules given by Eqs. (1), (2) and (4) have the same origin [18]. They are in fact a consequence of the Clifford algebra structure of the rest-frame supersymmetry algebra for massive and massless supermultiplets.

These sum rules can be summarized as follows: the supertrace of the p power of any operator O, which is bilinear in the rest-frame supersymmetry generators belonging to the SO(4N) algebra [SO(2N) in the massless case], satisfies the following property

$$\text{Supertrace } O^{p} = 0 \quad p < 2N \text{ } (p < N \text{ in the massless case}) \tag{5}$$

If we consider the helicity Λ in the massless case we get

$$\text{Supertrace } \Lambda^{p} = 0 \quad p < N \tag{6}$$

On one particle state this gives

$$\sum_{k=0}^{N}\binom{N}{k}(\lambda_{MAX} - \frac{k}{2})^{p}(-)^{k} = 0 \qquad p < N \tag{7}$$

This is nothing but formula (1). In the massless case we can combine the helicity Λ with an arbitrary SU(N) generator [contained in O(2N)] and we get

$$\text{Supertrace}\ \Lambda^h T^{i_1}_{j_1},\ldots,T^{i_n}_{j_n} = 0 \qquad h+n < N \tag{8}$$

For n = 2, by contraction, we obtain

$$\text{Supertrace}\ \Lambda^h C_2 = 0 \qquad h + 2 < N \tag{9}$$

which is Eq. (2).

Another interesting sum rule which involves the group theoretical factor of the SU(N) axial anomaly is

$$\text{Supertrace}\ \Lambda^h A = 0 \qquad h + 3 < N \tag{10}$$

where

$$A(R) = \frac{1}{2} \operatorname*{Tr}_{(R)}\ T^i_j\ \{T^j_k, T^k_i\} \tag{11}$$

Finally, in order to obtain (4) and its possible generalization we must apply (5) to the massive supersymmetry algebra with central charges and with maximal spin reduction. In this case we still have an SO(2N) algebra in the rest-frame but we apply Eq. (5) to the generators of $SU(2)_{spin} \times USp(N) \subset SO(2N)$. In spontaneously broken supergravity through dimensional reduction, the mass operator is an element of the Cartan subalgebra of USp(N) (which has rank N/2) so the particle masses depend on N/2 parameters. Since $\mathcal{M}$ belongs to USp(N) from (9) we get

$$\text{Superspace}\ (S^2)^h \mathcal{M}^{2k} = 0 \quad \text{for } 2h + 2k < N \quad \text{or} \quad h + k < [N/2] \tag{12}$$

S^2 in the spin Casimir.

For h = 0 we obtain the mass sum rules given by Eq. (4). For h > 0 Eq. (12) gives rise to new sum rules which combine spin factors and the masses.

It is evident that the previous sum rules, although they explain the vanishing of the one-loop β function, do not justify the vanishing of the β function at all higher loops as actually seems to be the case. It would be interesting to understand whether higher loop cancellations can be at all understood in the context of N extended supersymmetry by some factorization of group theoretical and kinematical factors occurring in multi-loop diagrams, which would eventually allow us to use a finite set of spin sum rules even at higher loops.

Finally, we would like to make some additional comment on a particular case of Eq. (4) which occurs in N = 1 supersymmetry theories, i.e., the mass relation [19]

$$\sum_J (-)^{2J} (2J + 1)\, m_J^2 = 0 \tag{13}$$

Equation (13) is in fact a functional identity in N = 1 renormalizable supersymmetric theories in the sense that Eq. (13) is fulfilled when the mass metric $m_J^2(z,z^*)$ for particles of spin j = 0, 1/2 and 1 is computed as a function of the physical scalar fields $z(z^*)$ of the theory. Equation (13) is true for any value of z, thus implying its validity also in spontaneously broken theories and in a class of explicitly broken theories with potential terms annihilated by the operator $\partial/\partial z^i \cdot \partial/\partial z^{*i}$. Equation (13) turns out to be the necessary and sufficient condition for the absence of quadratic divergences in spontaneously broken supersymmetric theories. The only exceptions to Eq. (13) are gauge theories with a chiral U(1) group (or many of them) such that the corresponding charge matrix Q is not traceless over the matter chiral multiplets. Under these circumstances Eq. (13) is then modified as follows

$$\sum_J (-)^{2J}(2J + 1)m_J^2(z,z^*) = \mathrm{Tr}\; Q\; D(z,z^*) \qquad (14)$$

where $D(z,z^*)$ is the auxiliary field of the corresponding U(1) gauge multiplet (A_μ,λ,D) and (14) is understood as a functional relation, valid for any value of z and z*.

Semi-realistic models considered in the literature [20,21] use an extra gauge factor $\tilde{U}(1)$ beyond SU(3) × SU(2) × U(1) in order to obtain a realistic spectrum for quarks, leptons and their scalar superpartners. Equation (14) shows that in order to avoid quadratic divergences this extra $\tilde{U}(1)$ group must be traceless. This implies that if ordinary matter has a positive $\tilde{U}(1)$ eigenvalue some additional fields and in particular coloured and charged fields must have negative $\tilde{U}(1)$ eigenvalues.

This fact has the potential danger of making problematic [22] the construction of a realistic model of electroweak and strong interactions with spontaneously broken supersymmetry, a realistic mass spectrum, the correct breaking of the gauge group $SU(3)_{colour} \times SU(2)_L \times U(1) \times \tilde{U}(1) \to SU(3)_{colour} \times U(1)_{em}$ and a $\tilde{U}(1)$ safe from A-B-J anomalies.

[1] E. Gildener and S. Weinberg, Phys. Rev. D13 (1976) 333;
S. Weinberg, Phys. Lett. 82B (1979) 387.

[2] L. Maiani, in Proc. of the Summer School of Gif-sur-Yvette page 3;
E. Witten, Nucl. Phys. B188 (1981) 513;
S. Dimopoulos and S. Raby, Nucl. Phys. B192 (1981) 353.

[3] J. Wess and B. Zumino, Phys. Lett. 49B (1974) 52;
J. Iliopoulos and B. Zumino, Nucl. Phys. B76 (1974) 310;
S. Ferrara, J. Iliopoulos and B. Zumino, Nucl. Phys. B77 (1974) 413.

[4] For a review see, P. van Nieuwenhuizen, Phys. Rep. 68, n.4 (1981) 189.

[5] E. Cremmer and B. Julia, Phys. Lett. 80B (1978) 48;
Nucl. Phys. B159 (1979) 41.

[6] For recent attempts at superunification see, e.g.:
J. Ellis, M.K. Gaillard and B. Zumino, LAPP-TH-44/ /CERN.TH.3152 (1981) to be published in Acta Physica Polonica;
S. Ferrara. CERN preprint TH.3158 (1981);
J.-P. Derendinger, S. Ferrara and C.A. Savoy, CERN preprint TH.3176 (1981).

[7] J. Wess and B. Zumino, Nucl. Phys. B78 (1974) 1;
S. Ferrara and O. Piguet, Nucl. Phys. B93 (1975) 261.

[8] M.T. Grisaru, M. Ročec and W. Siegel, Nucl. Phys. B159 (1979) 429;
M.T. Grisaru, Brandeis preprint, to appear in the Proceedings of the Supergravity School, Trieste (1981).

[9] O.V. Tarasov and A.V. Iadimirov, Dubna preprint (1980):
M.T. Grisaru, M. Ročec and W. Siegel, Phys. Rev. Lett. 45 (1980) 1063;
W.E. Caswell and D. Zanon, Phys. Lett. B100 (1980) 152.

[10] S. Ferrara and B. Zumino, unpublished (1979);
M. Sohnius and P. West, Phys. Lett. B100 (1981) 245.

[11] K. Stelle, Ecole Normale Sup. preprint LPTENS 81/24 (1981).

[12] P. Townsend and K. Stelle, to be published.

[13] E. Fradkin, Dubna preprint (1981).

[14] S.M. Christensen, M. Duff, G.W. Gibbons and M. Rocek, Phys. Rev. Lett. 45 (1980) 161.

[15] T.L. Curtright, Phys. Lett. 102B (1981) 17.

[16] J. Scherk and J. Schwarz, Nucl. Phys. B153 (1979) 61.

[17] E. Cremmer, J. Scherk and J. Schwarz, Phys. Lett. 84B (1979) 83;
S. Ferrara and B. Zumino, Phys. Lett. 86B (1979) 279.

[18] S. Ferrara, C.A. Savoy and L. Girardello, Phys. Lett. 105B (1981) 363.

[19] S. Ferrara, L. Girardello and F. Palumbo, Phys. Rev. D20 (1979) 403.

[20] P. Fayet in "Unification of Fundamental Particle Interactions", ed. by S. Ferrara, J. Ellis and P. van Nieuwenhuizen, (Plenum Press, N.Y., 1980).

[21] S. Weinberg, Harvard preprint HUTP-81/A047 (1981).

[22] R. Barbieri, S. Ferrara and D.V. Nanopoulos, CERN preprint TH.3226 (1982).

QUANTUM FIELD THEORY AND SPATIAL TOPOLOGY

C.J. Isham

ABSTRACT: The topology of physical three-space Σ influences significantly the canonical quantization of fields defined on Σ. We study in particular the relation between the Σ-topology and the topology of the canonical configuration space Q. Inequivalent quantum theories arise if $\Pi_1(Q)$ is non-trivial and constitute wide-ranging generalizations of the conventional Yang-Mills Θ-structure. We show how in a non-linear σ-model (resp. in Yang-Mills theory) Π_1 (Q) is equal to the group of homotopy classes of maps from the suspension of Σ into C/H (resp. Σ into G) and how this is determined by the cohomology groups of Σ. In the canonical quantization of gravity Π_1 (Q) is equal to the group of components of the diffeomorphism group of Σ.

1. INTRODUCTION

The small scale properties of spacetime are unknown and there is no a priori reason for supposing them to be trivial. A complete theory of quantum gravity may well involve physical spaces with complicated topologies and it has become important to understand the qualitative effects of such structures on otherwise conventional quantization methods. This applies both to quantum gravity itself and to the technically simpler subject of quantum field theory in a fixed background.

In quantum gravity it is not always possible to move freely from covariant to canonical approaches and an initial decision has to be made on whether to study three or four dimensional spaces. S.W. Hawking chose the latter and he and his collaborators have investigated the effects of spacetime topology on the functional integral quantization of gravity [1][2]. They use four dimensional riemannian (rather than pseudo-riemannian) manifolds with a wide selection of interesting topologies. Interpretational difficulties can arise and are related to the subtle problems involved in the analytic continuation to a physical pseudoriemannian spacetime.

The alternative, canonical, approach is discussed in this paper and we concentrate on the quantum effects produced by the topology of physical three-space Σ. A simple method of coding spatial topology into quantum field theory is to extend the concept of a field to include cross-sections of an arbitrary vector bundle over Σ. In the case of a real line bundle the sections have become known as 'twisted scalar fields' [3] or 'automorphic fields' [4] and quantum effects of the twists have been studied in detail [5].

A more subtle method of relating Σ-topology and quantum field theory arises when the canonical configuration space Q is itself topologically nontrivial. The topology of Q is closely related to that of Σ and is of particular importance if $\pi_1(Q) \neq 0$. A multiply connected configuration space generates a family of inequivalent quantizations of which the well known Yang-Mills θ-structure is a special case. We will study in particular the G/H-valued non linear σ-model (§3), Yang-Mills theory with an arbitrary internal symmetry group G (§4) and the canonical 'superspace' quantization of gravity (§5). The fundamental group $\pi_1(Q)$ in these three theories is shown to be $[S\Sigma, G/H]_*$, $[\Sigma, G]_*$ and $\pi_0(\mathrm{Diff}_*\Sigma)$ respectively where $S\Sigma$ is the suspension of Σ, $\mathrm{Diff}_*\Sigma$ is a certain subgroup of the full diffeomorphism group $\mathrm{Diff}\,\Sigma$ of Σ and

$[X,Y]_*$ denotes the set of homotopy classes of basepoint preserving maps from X into Y. The practical significance of these results becomes clearer if $\pi_1(Q)$ can be computed using only the topological properties of Σ. By using methods from algebraic topology, the sets $[S\Sigma, G/H]_*$ and $[\Sigma, G]_*$ can be related to the cohomology groups of Σ. This is important because the cohomology of a three-manifold Σ can be computed directly from $\pi_1(\Sigma)$ and, in three dimensions, the specification of Σ is essentially equivalent to fixing its fundamental group. The computation of $\pi_0(\mathrm{Diff}_*\Sigma)$ is much harder but fortunately there are several recent results relating the homotopy type of Diff Σ to the structure of Σ.

By these methods we can also show that a Gribov phenomenon occurs in both Yang-Mills and gravity theory and hence that it is impossible in either case to fix a global gauge.

The paper concludes with a brief introduction to some recent work on the full vacuum structure of a canonical Yang-Mills theory in a non trivial three-space .

2. NON-SIMPLY CONNECTED CONDIGURATION SPACES

Let us review briefly the quantum effects of a multiply connected configuration space Q [6-8] . In a path integral context let $K_{[p]}(q,q')$ denote the propagator obtained by summing over all the paths between q and q' that are homotopic to a particular path p. The full propagator is

$$K_\theta(q,q') = \sum_{[p]} \theta([p])\, K_{[p]}(q,q') \qquad (2.1)$$

where, for each $[p]$, $\theta([p])$ is a complex phase factor. Choose some arbitrary, but fixed, basepoint q_0 in Q and join q_0 to the points q and q' by a pair of paths ω and ω' respectively. Then the composite path $\omega'^{-1} v p v \omega$ defines an element of $\pi_1(Q,q_o)$, and (2.1) may be rewritten as

$$K_\theta(q, q') = \sum_\gamma \theta(\gamma)\, K_\gamma(q, q'); \; \gamma \in \pi_1(Q, q_o) \tag{2.2}$$

The element γ of $\pi_1(Q)$ depends on the choice of ω and ω' but, if θ is a character of $\pi_1(Q)$, the only effect of changing ω and/or ω' is to multiply $K_\theta(q,q')$ by a physically irrelevant phase factor. Hence the classifying set Θ of inequivalent quantizations is

$$\Theta = \mathrm{Hom}\,(\pi_1(Q),\, U1) \tag{2.3}$$

It is useful to rederive this result in a Hilbert space framework. In conventional canonical quantization the vector states are functions on Q and carry a representation of the canonical commutation relations in the (symbolic)form

$$(\hat{q}\psi)(q) = q\psi(q) \;\; ; \;\; (\hat{p}\psi)(q) = -i\hbar \frac{\delta\psi}{\delta q}(q) \tag{2.4}$$

The concept of a state may be generalised to include cross-sections of complex line bundles over Q. However, the associated principal U1-bundle cannot be arbitrary as this could entail the introduction of a connection which would change the dynamics radically. This connection can be gauged away locally on Q if the corresponding curvature two-form vanishes, which means that the bundle must be flat. Such bundles and connections are classified [9, 10] by the elements of $\mathrm{Hom}(\pi_1(Q),U1)$ and hence the above result is reproduced.

Discussion of the existence and effects of these characters has been mainly in the context of the quantization of a classical system with a finite dimensional configuration space Q whose multiple connectedness often derives from some physically unnatural constraints on the motion of the system. However we shall see that $\pi_1(Q) \neq 0$ is a natural occurence in several physically significant field theories and it becomes important to be able to compute $\pi_1(Q)$ using only the topological properties of the three-space Σ.

3. NON-LINEAR σ-MODELS

Non linear σ-models afford archetypal examples of dynamical systems with multiply connected configuration spaces. The canonical fields are maps from Σ (assumed compact) into the coset space G/H of two Lie groups G and H and hence carry a natural G-action. An invariant lagrangian is

$$\mathcal{L} = \frac{1}{2} \sqrt{g}\, g^{\mu\nu} \partial_\mu \phi^i \partial_\nu \phi^j \gamma_{ij}(\phi) \tag{3.1}$$

where γ_{ij} is a G-invariant metric on G/H. The configuration space is the set $Q = G/H_*^{\Sigma}$ of differentiable maps from Σ into G/H which, for convenience, are chosen to be basepoint preserving. Palais' theorems [11-12] demonstrate the homotopic equivalence of any topologies that are likely to be assigned to this function space and we can write unambiguously.

$$\pi_1(Q) = [S^1, G/H_*^{\Sigma}]_* = [S\Sigma, G/H]_* \tag{3.2}$$

where G/H_*^{Σ} carries the compact-open topology.

For practical purposes it is important to be able to compute $\pi_1(Q)$ in terms of explicit topological properties of Σ. A Postnikov decomposition of G/H will achieve this aim by relating $[S\Sigma, G/H]_*$ to the cohomology groups of Σ. [13-15]. For example consider the chiral case where $G = G_L x G_R$ and H is the diagonal subgroup. Then $G/H \approx H$ and we must study $[S\Sigma, H]_*$. As an example

(i) $H = SU(n)\ n \geq 3$

We have $\pi_1(SU(n)) = \pi_2(SU(n)) = \pi_4(SU(n)) = 0$

$$\pi_3(SU(n)) = Z$$

and hence, as far as maps from the 4-complex $S\Sigma$ into H are concerned, H may be replaced by (i.e. is 4-equivalent to) the Eilenberg-Maclane space K(Z,3). Therefore

$$\pi_1(Q) = [S\Sigma, SU(n)]_* = [S\Sigma, K(Z,3)]_*$$
$$= H^3(S\Sigma; Z) = H^2(\Sigma; Z) = H_1(\Sigma; Z) \quad (3.3)$$

and so the classifying set for quantum theories is

$$\underline{\textcircled{H} = \mathrm{Hom}(\pi_1(\Sigma), U1)} \quad (3.4)$$

Note that any compact orientable three-manifold Σ can be decomposed uniquely into a topological sum $\Sigma \approx \Sigma_1 * \Sigma_2 * \dots \Sigma_N$ of a finite set $\Sigma_1, \Sigma_2 \dots \Sigma_N$ of prime manifolds. However $\pi_1(\Sigma)$ is the free product

$$\pi_1(\Sigma) \approx \pi_1(\Sigma_1) * \pi_1(\Sigma_2) * \dots \pi_1(\Sigma_N)$$

and hence in evaluating $\textcircled{H}$ it suffices to know $\mathrm{Hom}(\pi_1(\Sigma_i), U1)$ for these prime factors.

(ii) $H = SU2$.

This case is more complicated than the previous one because $\pi_4(SU2) = Z_2$. A suitable Postnikov system is

$$\begin{array}{ccc} K(Z_2, 4) \longrightarrow (SU2)_4 & & \\ \downarrow & & \\ K(Z,3) & \xrightarrow{\alpha} & K(Z_2, 5) \end{array} \quad (3.5)$$

where $(SU2)_4$ is 4-equivalent to SU2. The classifying map χ defines an element of the cohomology group $H^5(K(Z,3); Z_2) \approx Z_2$ whose generator is Sq^2 (ι_3 mod 2), where ι_3 is the generator of $H^3(K(Z,3); Z) \approx Z$.

The Serre cohomology sequence of (3.5) with Z_2 coefficients shows that χ is this non trivial element. Mapping a 4-complex $\mathcal{M}$ into (3.5) yields the exact sequence

$$\to H^2(\mathcal{M}; Z) \xrightarrow{Sq^2 \circ \text{mod} 2} H^4(\mathcal{M}; Z_2) \to [\mathcal{M}, SU2]_* \to H^3(\mathcal{M}; Z) \to 0 \qquad (3.6)$$

which simplifies when $\mathcal{M} \approx S\Sigma$ since cup products vanish in the cohomology ring of a suspension. The result is

$$0 \to H^4(S\Sigma; Z_2) \to [S\Sigma, SU2]_* \to H^3(S\Sigma; Z) \to 0 \qquad (3.7)$$

i.e. $$0 \to H^3(\Sigma; Z_2) \to [S\Sigma, SU2]_* \to H^2(\Sigma; Z) \to 0 \qquad (3.8)$$

and the techniques of Larmour and Thomas [16] show that the short exact sequence (3.8) splits. Hence

$$\begin{aligned}[S\Sigma, SU2]_* &\approx H^3(\Sigma; Z_2) \oplus H^2(\Sigma; Z) \\ &= \quad Z_2 \quad \oplus H_1(\Sigma; Z)\end{aligned}$$

and the classifying set is

$$\underline{\text{Ⓗ} = Z_2 \oplus \text{Hom}(\pi_1(\Sigma); U1)} \qquad (3.9)$$

4. YANG-MILLS THEORY

In the canonical quantization of a Yang-Mills theory it is convenient to choose the gauge $A_0=0$. The canonical fields are then the spatial parts $\underline{A}(x)$ of the potential or, more precisely, the components of a connection in a principal G-bundle ξ over Σ. Such bundles are classified by elements of $H^2(\Sigma; \pi_1(G))$ but, for simplicity, we will restrict our attention to the trivial case. The gauge group $\mathcal{G}$ is then the set of differentiable functions from Σ to G and acts on the space of connections $\mathcal{C}$ by

$$A_i \rightarrow A_i^{\Omega} = \Omega A_i \Omega^{-1} + \Omega \partial_i \Omega^{-1} \qquad (4.1)$$

The physical configuration space Q is the set of equivalence classes $\mathcal{C}/\mathcal{G}$ and for quantum field theory purposes it is desirable that

i) Q is an infinite dimensional manifold

ii) $\pi_1(Q)$ is effectively computable in terms of the topological properties of Σ.

If $\mathcal{G}$ acts freely on $\mathcal{C}$ there is a chance that $\mathcal{C}$ could be a principal $\mathcal{G}$-bundle over $\mathcal{C}/\mathcal{G}$:

$$\begin{array}{ccc} \mathcal{G} & \rightarrow & \mathcal{C} \\ & & \downarrow \\ & & \mathcal{C}/\mathcal{G} \end{array} \qquad (4.2)$$

but in practice fixed points exist . These may be eliminated by restricting either $\mathcal{G}$ or $\mathcal{C}$ and for our purposes it is convenient to redefine $\mathcal{G}$ as the set of gauge functions that are equal to the identity $\mathbb{1}$ at the base point $\underline{x}_0$ of Σ [17]. Note that if $\underline{x}_0$ is the point at infinity of a one point compactification of a non compact manifold then $\mathcal{G}$ corresponds to the set of gauge functions that tend to $\mathbb{1}$ at large spatial distances. Under these circumstances the $\mathcal{G}$-action is free and it has been shown that $\mathcal{C}$ is a principal $\mathcal{G} = G_*^{\Sigma}$. bundle over the infinite differentiable manifold $Q = \mathcal{C}/\mathcal{G}$. [17-19] .

Now $\mathcal{C}$ is contractible and hence the homotopy exact sequence of this bundle gives

$$\pi_1(Q) = \pi_0(\mathcal{G}) = [\Sigma, G]_* \qquad (4.3)$$

By using a Postnikov decomposition of G we find that for most, simple, non-abelian, groups G[15],

$$\begin{aligned}[\Sigma, G]_* &= H^3(\Sigma; Z) \oplus H^1(\Sigma; \pi_1(G)) \\ &= Z \oplus \mathrm{Hom}(\pi_1(\Sigma), \pi_1(G))\end{aligned} \quad (4.4)$$

and in special cases (e.g. G=SO3) $[\Sigma, G]_*$ is a specific subgroup of (4.4). [15] [20] . Thus, typically,

$$\Theta = U1 \oplus \mathrm{Hom}(\mathrm{Hom}(\pi_1(\Sigma), \pi_1(G)), U1) \quad (4.5)$$

and we note that, when $\Sigma = S^3$, the usual classification of θ-states is obtained.

The justification for associating Θ with the conventional θ-structure lies in the assertion that cross-sections of a line bundle over Q, classified by $\theta \in \Theta$, are in bijective correspondence with complex valued functions Φ_θ on $\mathcal{C}$ satisfying

$$\Phi_\theta(\underline{A}^\Omega) = \theta([\Omega]^{-1})\, \Phi_\theta(\underline{A}) \quad (4.6)$$

which, in the special case $\Sigma = S^3$, is the familiar transformation of a θ-state.

In this treatment the θ-states are the basic entities and the (n,h)-states (the analogues of the 'n-states' of conventional $\Sigma = S^3$ Yang-Mills theory) are <u>defined</u> by

$$\Phi_{(n,h)}(\underline{A}) = \sum\!\!\!\!\!\!\int_{\theta \in \Theta} \theta^{-1}(n,h)\, \Phi_\theta(\underline{A}) \quad (4.7)$$

where the sum/integral is over the members of Θ .

Singer's main motivation for studying the principal $\mathcal{G}$-bundle (4.2) was to investigate the possibility of choosing global gauges [17]. These correspond to global cross-sections of the bundle and exist if and only if the bundle is trivial. If $\mathcal{C} \simeq \mathcal{G} \times Q$ then $\pi_n(\mathcal{G}) = 0$ for all n; hence the existence of a single non-vanishing homotopy group of $\mathcal{G}$ is sufficient to create the Gribov phenomenon (the non existence of a gauge). Since, in particular, $\pi_0(\mathcal{G}) = [\Sigma, G]_*$, eqn (4.4) shows that the Gribov effect arises for any non abelian group G. If $G = \overset{n}{\underset{i=1}{\times}} U1$ say, then $[\Sigma, G]_* = \overset{n}{\underset{i=1}{\oplus}} H^1(\Sigma; Z)$ and a non vanishing first Betti number of Σ will produce the Gribov phenomenon.

5. THE CANONICAL QUANTIZATION OF GRAVITY

In the canonical quantization of gravity the basic field variables are the riemannian metric g_{ij} on Σ and its canonical conjugate π^{kl}. General coordinate invariance produces first class constraints,

$$\mathcal{H}_\perp(g, \pi) = 0 \tag{5.1}$$

$$\mathcal{H}_i(g, \pi) = 0 \qquad i = 1, 2, 3. \tag{5.2}$$

where $\mathcal{H}_\perp$ and $\mathcal{H}_i$ are respectively the generators of deformations normal to Σ and diffeomorphisms of Σ. The vector states are functionals $\Psi(g)$ and (5.2) is usually taken to imply that Ψ is invariant under the group of spatial diffeomorphisms Diff Σ and hence can be regarded as a function on Wheeler's superspace [21,22] - the set of equivalence classes Riem Σ/Diff Σ of the Diff Σ action on the space Riem Σ of riemannian metrics,

$$\begin{array}{ccc} \text{Diff}\,\Sigma & \longrightarrow & \text{Riem}\,\Sigma \\ & & \downarrow \\ & & \mathcal{S}(\Sigma) = \text{Riem}\,\Sigma/\text{Diff}\,\Sigma \end{array} \tag{5.3}$$

Guided by the Yang-Mills theory we extend the family of vector states to include cross-sections of arbitrary flat complex line bundles over $\mathcal{S}(\Sigma)$ and then study the group $\pi_1(\mathcal{S}(\Sigma))$. Unfortunately the Diff Σ action on Riem Σ has fixed points - the metrics with isometries. Consequently $\pi_1(\mathcal{S}(\Sigma))$ is difficult to compute and $\mathcal{S}(\Sigma)$ is not a differentiable manifold (it is stratified; Fischer [23]). One resolution of this problem is to restrict the diffeomorphisms to those that leave fixed the base point and its tangent space. The resulting group $\mathrm{Diff}_*\Sigma$ does act freely on Riem Σ and the work of Ebin [24] and Fischer [23] shows that $\mathrm{Diff}_*\Sigma \to \mathrm{Riem}\,\Sigma \to \mathcal{S}_*(\Sigma)$ is a principal $\mathrm{Diff}_*\Sigma$ bundle. B. De Witt has called the quotient space $\mathcal{S}_*(\Sigma) = \mathrm{Riem}\Sigma/\mathrm{Diff}_*\Sigma$ "extended superspace". As desired, $\mathcal{S}_*(\Sigma)$ is a genuine infinite dimensional manifold and, since Riem Σ is contractible, $\pi_1(\mathcal{S}_*(\Sigma))$ is given by

$$\pi_1(\mathcal{S}_*(\Sigma)) = \pi_0(\mathrm{Diff}_*\Sigma) \tag{5.4}$$

and therefore

$$\boxed{H} = \mathrm{Hom}(\pi_0(\mathrm{Diff}_*\Sigma), U1) \tag{5.5}$$

With the exception of $\Sigma = S^3$, $\pi_0(\mathrm{Diff}_*\Sigma)$ seems to be non-trivial and hence a θ-structure and a Gribov effect both exist. For further details see [25-26].

A number of extensions of this work are possible:

(i) We could use conformal superspace $\mathcal{J}_*(\Sigma)$ where one of the three degrees of freedom per space point in $\mathcal{S}(\Sigma)$ has been identified as an intrinsic time and factored out. This would enable the time evolution problem to be studied by fixing a time gauge and solving (5.1) classically. There should then be gravitational analogues of the tunnelling instantons of Yang-Mills theory (NB. These should not be confused with Hawking's gravitational instantons [1][2]).

(2) If $\mathcal{S}(\Sigma)$ is employed, Teitelboim's functional integral[27] approach to canonical quantization could be used to look for tunnelling phenomena.

(3) A combined Yang Mills/quantum gravity theory could be studied with the gauge group $\mathcal{G}$ now being the set of morphisms of the principal G-bundle ξ. If ξ is nontrivial (i.e. it corresponds to a non zero characteristic class in $H^2(\Sigma;\pi_1(G))$ the problem of selecting a freely acting subgroup of $\mathcal{G}$ is modified because certain diffeomorphisms of Σ, possibly including some isometries, will not lift to a automorphism of ξ.

(4) A similar theory applies to supergravity and the homotopy properties of the supergauge group become of interest. The global properties of this group have not received much attention.

(5) The interesting work of Friedman and Sorkin [28] on the existence of SO3 spin-$\frac{1}{2}$ representations on the quantum gravity state space can be extended to the present case where Σ is compact. A 2π rotation is performed in some neighbourhood of the basepoint and then extended to a global diffeomorphism R of Σ. If this diffeomorphism is not homotopic to the identity in $\mathrm{Diff}_*\Sigma$ there may exist multivalued wave functions Ψ (i.e. functionals on the covering space of $\mathcal{S}(\Sigma)$) such that $\hat{R}\Psi = -\Psi$ where $(\hat{R}\Psi)(g) := \Psi(R^{-1*}g)$. Such transformations indicate the presence of spin-$\frac{1}{2}$.

(6) This technique could be extended to the examples (3) and (4) above. In particular a $SU3/Z_3$ Yang-Mills theory coupled to gravity might, if the topology of Σ was suitable, yield states that carry SO3 spin-$\frac{1}{2}$ and internal $SU3/Z_3$ triality non-zero representations. Hence, with respect to both their spatial spin and internal symmetry properties, quarks could be regarded as topological excitations of Yang-Mills and

gravity fields.

6. YANG-MILLS VACUA

The discussion in §4 does not exhaust the topological content of a canonically quantized Yang-Mills theory. General (n,h)-states are classified by elements of $[\Sigma, G]_* = Z \oplus Hom(\pi_1(\Sigma), \pi_1(G))$ but the vacuum states have an additional structure. The (n,h)-vacua are customarily regarded as wave functionals $\Psi(A)$ peaked around classical solutions to the zero energy condition

$$F_{ij} = 0 \tag{6.1}$$

If $\pi_1(\Sigma) = 0$, the only solution is

$$A_i = \Omega \partial_i \Omega^{-1} \quad , \quad \Omega : \Sigma \to G \tag{6.2}$$

which leads to a classification of vacua by the homotopy classes of the gauge functions, i.e. by $[\Sigma, G]_*$. However if $\pi_1(\Sigma) \neq 0$ then there exist solutions to (6.1) that are generalizations of (6.2) in the form

$$A_i(\underline{x}) = D(\underline{y}) \partial_i D(\underline{y})^{-1} \tag{6.3}$$

where $D(\underline{y})$ is a G-valued function on the universal covering space $\hat{\Sigma}$ of Σ satisfying $D(\underline{y}\gamma) = D(\underline{y}) h(\gamma)$ with $\underline{y} \in \hat{\Sigma}$ being any covering point of $\underline{x}$ and h is any homomorphism of $\pi_1(\Sigma)$ into G. If $\mathcal{D}$ denotes the set of all such functions then the analogues of n-vacua are classified by elements of $\pi_0(\mathcal{D})$.

It is shown in [29-30] that the map $\mathcal{D} \to R$, $D \rightsquigarrow h$ (h is the homomorphism in $D(\underline{y}\gamma) = D(\underline{y})h(\gamma)$) defines a principal G_*^Σ bundle over the space R of homomorphisms from $\pi_1(\Sigma)$ to G that induce G-bundles on

Σ that, qua G-bundles, are trivial. The classifying set $\pi_0(\mathcal{D})$ is then enclosed in the exact sequence

$$\rightarrow \pi_1(\mathcal{R}) \rightarrow [\Sigma, G]_* \rightarrow \pi_0(\mathcal{D}) \rightarrow \pi_0(\mathcal{R}) \rightarrow *$$

and can be studied as such. Full details may be found in [29-30] and in G. Kunstatter's article in this volume.

REFERENCES

[1] HAWKING, S.W., Spacetime foam, Nuc. Phys. B144 (1978) 349.

[2] HAWKING, S.W., The path integral approach to quantum gravity. In General Relativity - an Einstein centenary survey, eds. S.W. Hawking and W. Israel. Cambridge University Press (1979) 746.

[3] ISHAM, C.J., Twisted quantum fields in a curved spacetime, Proc. Roy. Soc. A362 (1978) 383.

[4] DOWKER, J.S., BANACH, R., Automorphic field theory - some mathematical issues, Jour. Phys. A12 (1979) 2527.

[5] See for example the references cited in ISHAM C.J., Quantum Gravity - An Overview. In Quantum Gravity - A second Oxford symposium, eds. C.J. Isham, R. Penrose and D.W. Sciama. Oxford University Press (1981).

[6] SCHULMAN, L.S. Approximate topologies, Jour. Math. Phys. 12 (1971) 304.

[7] LAIDLAW, M., DEWITT C. Feynman functional integrals for systems of indistinguisbale particles, Phys. Rev. D3 (1971) 1375.

[8] DOWKER, J.S., Quantum mechanics and field theory on multiply connected and on homogeneous spaces. Journ. Phys. A5 (1972) 936.

[9] MILNOR, J., On the existence of a connection with curvature zero. Comm. Math. Helv. 32 (1957) 215.

[10] KOSTANT, B., Quantization and unitary representations. In Lecture notes in mathematics, Vol.170 Springer Verlag (1970).

[11] PALAIS, R.S., Homotopy theory of infinite dimensional manifolds, Topology 5 (1966) 1-16.

[12] PALAIS, R.S., Foundations of global non linear analysis. Benjamin (1968).

[13] THOMAS, E., Seminar on fiber spaces, Lecture notes in mathematics Vol. 13, Springer Verlag (1966).

[14] AVIS, S.J., ISHAM C.J., Quantum field theory and fibre bundles in a general spacetime. In Recent developments in gravitation. Cargese 1978. Eds. M. Levy and S. Deser, Plenum Press (1979).

[15] ISHAM C.J., Vacuum tunnelling in static spacetimes. In Essays in honour of Wolfgang Yourgrau, ed. A. van der Merwe, Plenum Press (1982).

[16] LARMORE, L.L., THOMAS, E., Group extensions and principal fibrations, Math. Scand. 30 (1972) 227.

[17] SINGER, I., Some remarks on the Gribov ambiguity. Comm. Math. Phys. 60 (1978) 7.

[18] MITTER, P.K., VIALLET, C.M., On the bundle of connections and the gauge orbit manifold in Yang-Mills theory, Comm. Math. Phys. 79, (1981) 457.

[19] NARISIMHAN, M.S., RAMADAS T.R., Geometry of SU2 gauge fields Comm. Math. Phys. 67 (1979) 121.

[20] SHASTRI, A.R., WILLIAMS, J.G., ZVENGROWSKI, P., Kinks in general relativity. Int. Jour. Theor. Phys. 19 (1980) 1.

[21] WHEELER, J.A., Geometrodynamics and the issue of the final state. In Relativity, groups and topology, eds. C. DeWitt and B.S. DeWitt, Blackie (1963).

[22] WHEELER, J.A. Superspace and the nature of quantum geometrodynamics. In Battelle Rencontres 1967, eds. C. DeWitt and J.A. Wheeler. Benjamin (1968).

[23] FISCHER, A.E., The theory of superspace, In Relativity, eds. M. Carmeli, S. Fickler and L. Witten, Plenum (1967).

[24] EBIN, D.G., The manifold of riemannian metrics. Proc. Symp. Pure Math. 15 (1970) 11.

[25] ISHAM, C.J., Topological θ-sectors in canonically quantized gravity, Phys. Letts B106 (1981) 188.

[26] ISHAM, C.J., θ-states induced by the diffeomorphism group in canonically quantized gravity. In The quantum structure of space and time, eds. M.J. Duff and C.J. Isham. Cambridge University Press (1982).

[27] TEITELBOIM, C., Quantum mechanics of the gravitational field. Austin preprint (1981).

[28] FRIEDMAN, J.L., SORKIN R.D., Spin-$\frac{1}{2}$ from gravity, Phys. Rev. Lett.

[29] ISHAM, C.J. KUNSTATTER, G. Yang-Mills canonical vacuum structure in a general three-space. Phys. Letts B102 (1981) 417.

[30] ISHAM, C.J., KUNSTATTER, G. Spatial topology and Yang-Mills vacua, Journ. Math. Phys. (1982).

GAUGE FIELD THEORIES AND THE EQUIVALENCE PRINCIPLE

C.A. Orzalesi

1. Introduction

Following the works of Utiyama, Kibble and Sciama, it is often emphasized that general relativity can be based on a gauge principle, a fact which makes gravity theory similar to gauge theories of the Yang-Mills type. One of my purposes here is to reverse this emphasis, by showing that the theory of interacting gravitational and Yang-Mills fields originates from general relativity on a suitably "extended" spacetime. Thus, as it happens with ordinary general relativity, also Yang-Mills gauge theories can be based on an equivalence principle, whereby the fundamental geometric origin of gauge theories becomes manifest.

The generalized equivalence principle cannot be formulated on ordinary spacetime, but it requires that the theory be a dynamical theory of the geometry of a multidimensional spacetime having the structure of a differentiable fibre bundle. From the geometrical multidimensional theory, the usual theory is recovered by dimensional reduction. The point-particle equations for Yang-Mills charges follow directly from the equivalence principle. A prediction is that point particles with nonzero Yang-Mills charge necessarily have a nonvanishing mass.

The multidimensional action principle, combined with the requirement that the multidimensional vacuum be "strongly" gauge-invariant, leads to another prediction: the cosmological constant for gravity in ordinary spacetime must vanish identically.

Dimensional extension and the fibre-fundle aspects of extended spacetime are reviewed in Section 2. The group-averaging technique of dimensional reduction and the notion of geometrodynamical theories are discussed in Section 3.

In Section 4, we derive and discuss the geodesic equations, which are used in the generalized equivalence principle. The issues of gauge invariance and of the cosmological constant are discussed in Section 5.

2. Dimensional Extension

Fibre bundles occur naturally whenever gauge groups are involved, see e.g. Ref. [1].

Suppose that a theory on ordinary spacetime M^4 has local dynamic variables $y_i(x)$ and a local gauge Lie group G with N parameters. By this, I mean that, for each $x \in M^4$ and $X \in G$, we have assigned a transformation law, $y_i(x) \xrightarrow[(x,X)]{} y_i^{X(x)}(x) = D_x(X) \cdot y_i(x)$, satisfying the group properties. Thus, at each $x \in M^4$ we have a representation $D_x(G)$ of G, and we may as well say that at each x we have an exact copy G_x of G: G_x is then represented as a group of transformations on the dynamic variables at x. Note that, at this point, it is by no means necessary that G be a symmetry group of the dynamical equations. Now, it is natural to put all the G_x's together and to consider the extended spacetime, $M^{4+N} = \bigcup_{x \in M^4} G_x$. Locally, M^{4+N} looks like the product $M^4 \times G$, and M^4 can be recovered as a space of gauge orbits, $M^4 = M^{4+N}/G$. Globally, M^{4+N} has the structure of a principal fibre bundle[2] $M^{4+N}(M^4,G)$ with base M^4, structure group G and projection π: $M^{4+N} \longrightarrow M^4 = M^{4+N}/G$. This is a very simple structure: attached to each $x \in M^4$ we have a fibre $\pi^{-1}(x) = G_x$, and an isomorphism σ_x: $G_x \longrightarrow G$ by $\sigma_x(\xi) = X$, $\xi \in G_x$, $X \in G$. A right action of G on each G_x (and hence on all of M^{4+N}) is defined by $R_Y\xi \equiv \xi \cdot Y \equiv \sigma_x^{-1}(X \cdot Y)$, where $\xi \in G_x, X, Y \in G$, so that also $X \cdot Y \in G$ and $R_Y\xi \in G_x$.

The differential $d\sigma_x^{-1}$ maps (isomorphically) vectors $T|_X$ tangent to G at X over vectors $V|_\xi$ tangent to G_x at ξ. By varying x over M^4 and X over G, we obtain a vector field V over M^{4+N}, such that $d\sigma_x V|_\xi = T|_X$. In particular, from a basis T_A, $A = 4 + 1, \ldots, 4 + N$ of the Lie algebra g of G, in this way we create a set of N fundamental vector fields V_A, $A = 4 + 1, \ldots, 4 + N$, over M^{4+N}. The T_A's are left-invariant: if L_Y: $X \longrightarrow Y \cdot X$ denotes left translations of G, then $T_A|_{Y \cdot X} = dL_Y \cdot T_A|_X$; therefore, $T_A|_X$ is determined at each $X \in G$ by left translation from $T_A|_I$, where I is the identity of G. The T_A's generate right translations on G: $X^A + \delta Y^B T_B(X^A)$ are the coordinates of $X \cdot \delta Y$ with δY infinitesimal. Similarly, the V_A's are left-invariant and generate right translations along each fibre.

A vector tangent to fibres is called vertical; the V_A's form a complete set of vertical vector fields, and the linear span $\mathcal{V}|_\xi$ of the $V_A|_\xi$ forms the N-dimensional vertical subspace of the tangent space $T_\xi M^{4+N}$. Thus, if V is vertical at ξ, then $V|_\xi = c^A(\xi) V_A|_\xi$, and $V|_\xi$ is tangent at ξ to the fibre G_x through ξ, where $x = \pi(\xi)$.

We can add to $\mathcal{V}|_\xi$ a subspace $\mathcal{H}'|_\xi$ of $T_\xi M^{4+N}$ such that $\mathcal{V}|_\xi + \mathcal{H}'|_\xi = T_\xi M^{4+N}$; this can be done in many ways, and in general $\mathcal{V}|_\xi \cap \mathcal{H}'|_\xi$ is not empty. A bundle connection is a (smooth) decomposition, for each $\xi \in M^{4+N}$, $T_\xi M^{4+N} = \mathcal{V}|_\xi + \mathcal{H}|_\xi$, where

$V_{|\xi} \cap H_{|\xi}$ is empty and $H_{|\xi\cdot Y} = dR_Y \cdot H_{|\xi}$.

Thus, a bundle connection is a smooth right-invariant direct sum decomposition of $T_\xi M^{4+N}$ into the vertical subspace $V_{|\xi}$ and a horizontal subspace $H_{|\xi}$. Given a bundle connection, any vector $X_{|\xi} \in T_\xi M^{4+N}$ is uniquely decomposed as

$$X_{|\xi} = \mathrm{hor}X_{|\xi} + \mathrm{Ver}X_{|\xi}, \text{ where } \mathrm{hor}X_{|\xi} \in H_{|\xi}, \mathrm{Ver}X_{|\xi} \in V_{|\xi}. \quad (2.1)$$

To visualize the structure of M^{4+N}, one can use adapted local coordinates $(\xi^\alpha) = (x^a, X^A)$, where x^a, $a = 1,\ldots,4$ are coordinates for $x = \pi(\xi)$ and X^A, $A = 4 + 1,\ldots,4 + N$ are coordinates for $X = \sigma_x(\xi) \in G$. Thus the projection $x = \pi(\xi)$ has coordinates $\pi^a(\xi) = x^a$. One can think of M^4 as being immersed in M^{4+N} as the submanifold (x,I); in local coordinates, with $X^A = 0$ the coordinates of $I \in G$, M^4 is then locally described as the submanifold $(x^a,0)$ of M^{4+N}.

For a smooth vector field $X = X^\alpha \partial/\partial\xi^\alpha$ on M^{4+N}, the projection gives $d\pi \cdot X = X^a \partial/\partial x^a$, which is a (not necessarily smooth) vector field on M^4. Clearly, $d\pi \cdot \mathrm{Ver}X = 0$, so that $d\pi \cdot X = d\pi \cdot \mathrm{hor}X$, and there are ∞^N vectors at ξ which have the same "shadow" $h_{|x} = h^a \partial_{a|x}$ at $x \in M^4$. But among these, there is only one which has vanishing vertical component: this unique vector at ξ, which projects onto $h_{|x}$ and is horizontal, is called the (horizontal) lift $\hat{h}_{|\xi}$ of $h_{|x}$ from $x \in M^4$ to $\xi \in M^{4+N}$.

Notation: we will continue to use Greek, lower-case Latin and capital Latin symbols and indices to refer respectively to M^{4+N}, to M^4 or to horizontal directions and to G or to vertical directions.

We can form a field of reference frames on M^{4+N} as follows: we take the lift ($\hat{h}_a$, $a = 1,\ldots,4$) of a field of reference frames (h_a) on M^4, and we combine this with the fundamental vector fields, so that $(\kappa_\alpha) = (\hat{h}_a, V_A)$ is our field of reference frames on M^{4+N}. Since the $\hat{h}_a$'s are horizontal and the V_A's are vertical, such frames are adapted to the bundle connection. Note also that the lift construction implies that $\hat{h}_{a|\xi\cdot Y}$ is obtained by right translation from $\hat{h}_{a|\xi}$, i.e. the $\hat{h}_a$'s are right-invariant. For such adapted frames, the vertical Lie derivatives of $\hat{h}_a$ vanish:

$$L_{V_A} \hat{h}_a = [V_A, \hat{h}_a] = 0 \quad , \quad (2.2)$$

while the V_A's clearly represent the Lie algebra $\mathcal{g}$ of G:

$$[V_A, V_B] = C^D_{\cdot AB} V_D \quad , \tag{2.3}$$

where $C^D_{\cdot AB}$ are the structure constants of $\mathcal{g}$ in the basis (T_A).

In particular, for the horizontal lift frame $(\hat{\partial}_a, V_A)$, where $\hat{\partial}_a$ is the lift of $\partial_a = \partial/\partial x^a$, one has $d\pi\cdot[\hat{\partial}_a,\hat{\partial}_b] = [d\pi\cdot\hat{\partial}_a, d\pi\cdot\hat{\partial}_b] = [\partial_a,\partial_b] = 0$, so that $[\hat{\partial}_a,\hat{\partial}_b]$ is purely vertical and can be expanded along the V_A's, with ξ-dependent coefficients $\mathcal{F}^A_{ab}$:

$$[\hat{\partial}_a, \hat{\partial}_b] = \mathcal{F}^A_{ab}\, \mathcal{V}_A \ . \tag{2.4}$$

To further understand (2.4), note that ∂_a is a vector on M^4: hence, it has 4 components; on the other hand, $\hat{\partial}_a$ is a vector on M^{4+N} and hence it has (4+N) components. Thus, in general one has

$$\hat{\partial}_a = \partial_a - \mathcal{A}^A_a V_A \quad , \tag{2.5}$$

with nonvanishing $\mathcal{A}^A_a$'s. In other words, in general $\partial/\partial\xi^a$ is not horizontal in a given bundle connection and, indeed, the $\mathcal{A}^A_a$ are coefficients which specify the connection in terms of the frame (∂_a, V_A).

From (2.2) one finds that the X-dependence of $\mathcal{A}^A_a(\xi)$ is determined as follows[3]:

$$\mathcal{A}^A_a(\xi) = \mathcal{A}^A_a(x,X) = A^B_a(x) D^{-1}{}^A_B(X) \quad , \tag{2.6}$$

where $D^{-1}{}^A_B$ is the matrix of the inverse adjoint representation of G in the coordinates (X^A). Then (2.4)-(2.6) give

$$\mathcal{F}^A_{ab}(x,X) = F^B_{ab}(x) D^{-1}{}^A_B(X) \ , \ F^B_{ab}(x) = A^B_{[b,a]} + C^B_{\cdot CD} A^C_a A^D_b \ . \tag{2.7}$$

We see that, from the lift $(\hat{\partial}_a)$ of (∂_a), 4·N new functions $A^B_a(x)$ are generated on M^4; these are the usual Yang-Mills potentials for the gauge group G, and F^B_{ab} is the corresponding field strength. If one performs a change of fibre coordinates,[3]

$$x^a \to x'^a = x^a \ , \ X^A \to X'^A = X^A + \delta X^A(x) \ , \ X' = Y(x)\cdot X \tag{2.8}$$

then the vector transformation law for the natural-basis components $\hat{h}^\alpha_a$ of $\hat{h}_a$ implies that the A^B_a's and F^B_{ab}'s undergo the familiar Yang-Mills gauge transformations. Correspondingly, under (2.8) one is changing[1,4] the identification of M^4 as a four-dimensional section in M^{4+N}.

3. Geometrodynamical Gauge Theories

A more concrete definition of "horizontality" can be achieved through the use of a metric on M^{4+N}.

To motivate this extension of the metric, let $\underset{\sim}{g}$ be an Einstein metric on M^4. Here we take G to be an internal symmetry group; since the graviton is neutral, $\underset{\sim}{g}$ is invariant under G. Further, we assume that G is a metrizable group, in particular a compact Lie group. Then, given a right-G-invariant metric $\underset{\sim}{G}$ on G,

$$G_{AB}(X) = (T_A \cdot T_B)|_X \ , \quad L_{T_A}\underset{\sim}{G} = 0 \ , \tag{3.1}$$

we can use the isomorphisms σ_x^{-1} and the lift to induce a non-degenerate metric γ on M^{4+N}, and one can use γ to simply identify "horizontality" with γ-orthogonality to the fibres. To be consistent with the same notion in terms of a bundle connection, γ must be right-G-invariant, i.e. the V_A's must be Killing vector fields for $\underset{\sim}{\gamma}$:

$$L_{V_A}\underset{\sim}{\gamma} = 0 \ , \quad A = 4+1,\ldots, 4+N \ . \tag{3.2}$$

Conversely, given a metric γ on M^{4+N} satisfying (3.2), a bundle connection is defined through γ-orthogonality. Actually, the following stronger property holds: if γ admits N complete Killing vector fields forming the Lie algebra g of a compact G, then M^{4+N} has the principal fibre bundle structure $M^{4+N}(M^4, G)$.

The property (3.2) constrains the X-dependence of $\underset{\sim}{\gamma}$: one finds[3]

$$\hat{\gamma}_{ab} = \hat{\gamma}_{ab}(x) \ , \ \hat{\gamma}_{aA} = 0 \ , \ \hat{\gamma}_{AB} = \phi_{EF}(x) D^{-1}{}^E{}_A(X) D^{-1}{}^F{}_B(X) \ . \tag{3.3}$$

Thus, although M^4 and G need not be isometrically imbedded into M^{4+N}, the form of the X-dependence of $\underset{\sim}{\gamma}$ is essentially fixed by (3.2).

Hereafter, we only consider bundle connections defined in terms of a right G-invariant metric γ on M^{4+N}. In all cases of interest, M^4 is conformally immersed in M^{4+N}, in the sense that the Einstein metric g on M^4 satisfies $\hat{\gamma}_{ab} = \Phi g_{ab}$; the conformal factor Φ is, in practice, determined from $\det\phi_{EF}$, see below.

Having a concrete (metric) visualization of the bundle connection, we now return to the connection coefficients $\mathcal{A}^B_a$ and associated field strengths $\mathcal{F}^B_{ab}$. We can regard the corresponding Yang-Mills quantities, $A^B_a(x)$ and $F^B_{ab}(x)$, as the "shadows" on M^4 of $\mathcal{A}^B_a(\xi)$ and $\mathcal{F}^B_{ab}(\xi)$; indeed, from (2.9-10) we have $A^B_a(x) = \mathcal{A}^B_a(x,I)$; $F^B_{ab}(x) = \mathcal{F}^B_{ab}(x,I)$.

Note that $A^B_a(x)$ and $F^B_{ab}(x)$ cannot be regarded as geometric quantities on M^4, since M^4 knows nothing about group indices. On the other hand, $\mathcal{A}^A_a(\xi)$ and $\mathcal{F}^A_{ab}(\xi)$ do have geometric meaning on M^{4+N}. Similarly, gauge-transformation properties have no geometric meaning on M^4, but they are the 4-shadows of corresponding tensorial coordinate transformation laws on M^{4+N}.

The above motivates the following definition: a (property of a) quantity $y^{AB\cdots}_{ab\cdots}(x)$ on M^4 has geometric origin if it is the shadow on M^4 of a corresponding geometric (property of a) quantity $\gamma^{\gamma\delta\cdots}_{\alpha\beta\cdots}(\xi)$ on M^{4+N}.

Next, we consider dynamical theories on M^{4+N} and on M^4. Let $L_{4+N}(\gamma,\cdots) = L_{4+N}[\xi]$ be a Lagrangian for a theory on M^{4+N}:

$$\delta\theta_{4+N} \equiv \delta\int_{M^{4+N}} L_{4+N}\,[\xi]d^{4+N}\xi = 0 \,. \tag{3.4}$$

Suppose that, for some reason, we only consider those solutions of (3.4) such that the resulting X-dependence of $L_{4+N}[x,X]$ is "frozen" to a preassigned form. Then, we can consider a dimensional reduction of the theory (3.4), obtained by performing the d^NX-integral before effecting the variations δ. In this way, we obtain a theory on M^4:

$$\delta A_4 = \delta\int_{M^4} L_4\,[x]d^4x = 0 \quad , \tag{3.5}$$

where the reduced Lagrangian is

$$L_4 \equiv \int_{G_x} L_{4+N}\,[x,X]d^NX \,. \tag{3.6}$$

The theory (3.5-6) is an effective theory of G-averages associated to the theory (3.4). As such, the theory (3.5-6) has fewer equations than the full theory (3.4).

Other procedures of dimensional reduction are known: see e.g., ref. [5].

In general, the theories (3.4) and (3.5-6) are inequivalent: the theory (3.4) is richer than (3.5-6), and we will make a use of the additional ("vertical") equations resulting from (3.4).

A recurring theme of this lecture is that properties, which are not geometric on M^4, can originate from the geometry of M^{4+N}. This can happen also for the dynamics, and it motivates the following definition:

A theory on M^4 with Lagrangian L_4 and gauge group G has <u>geometric origin</u> if L_4 is obtained by dimensional reduction (as in (3.5)) from a theory on M^{4+N}, with a Lagrangian L_{4+N} which is a geometric object on M^{4+N}.

The standard example of a geometrodynamical theory on M^{4+N} is the Kaluza-Klein-Jordan-Thiry-Lichnerowicz unified theory of gravitation and electromagnetism[6-9], generalized to a nonabelian compact gauge group[10-13,4,3].

The basic hypotheses of such multidimensional unified theories are as follows:

<u>i</u>. The Einstein-Hilbert action principle on M^{4+N}:

$$\delta\Theta_{4+N} = -\delta\int_{M^{4+N}} (R(\{\})-\lambda)|\gamma|^{\frac{1}{2}}\, d^{4+N}\xi = 0 \quad , \tag{3.7}$$

where $\{\}$ denotes the Levi-Civita connection for the metric $\underset{\sim}{\gamma}$, $R(\{\})$ is the corresponding Riemann-Christoffel curvature scalar, and λ is a cosmological constant.

<u>ii</u>. The existence of N complete Killing vector fields for $\underset{\sim}{\gamma}$ (eq. (3.2)), faithfully representing (eq. (2.3)) the Lie algebra $\mathcal{g}$ of a compact N-parameter Lie group G.

As already remarked, <u>ii</u> implies the principal fibre bundle structure $M^{4+N}(M^4,G)$. To make contact with a theory on M^4, one usually assumes that the dimensional reduction of (3.7) should lead --- through a conformal transformation --- to the Einstein term $R^{(4)}(\{\})$ in the 4-Lagrangian. This fixes[14] the conformal factor, $\Phi = |\det \phi_{EF}|^{-\frac{1}{2}}$, and (3.3) becomes

$$\hat{\gamma}_{ab} = \Phi(x) g_{ab}(x) \quad , \quad \hat{\gamma}_{aA} = 0 \quad , \quad \hat{\gamma}_{AB} = \phi_{EF}(x) D^{-1E}{}_{A}(X) D^{-1F}{}_{B}(X) . \tag{3.8}$$

The Levi-Civita connection of (3.8) is easily computed[3,12] as well as the associated curvature, and the dimensional reduction (3.5-6) can be carried out.

The resulting theory on M^4 describes[4,14,3] the interaction of ordinary gravity with a coupled system of Yang-Mills fields and vertical metric fields ϕ_{EF} having a role of x-dependent Yang-Mills coupling "constants", with a cosmological term $\bar{\lambda}$.

One can further restrict (3.8) by imposing that the isomorphisms o_x^{-1} be x-independent isometries for a bi-invariant Killing-Cartan metric on G; this amounts to the assumption that, for G simple,

$$\hat{\gamma}_{AB} = -c^2 \delta_{AB} \qquad (3.9)$$

in an orthogonal basis (T_A) for g. Eq. (3.9) is the Kaluza-Klein constraint (γ_{55} = constant) generalized to a nonabelian G. With (3.9), the dimensionally reduced theory describes the Einstein-Yang-Mills coupled system, with a cosmological constant $\sim\lambda - R^{(G)}(\{\})$; here, $R^{(G)}(\{\})$ is the Riemann curvature of G for the Levi-Civita connection on the group G.

Disregarding for the moment the cosmological constant, we see that the Einstein-Yang-Mills theory on M^4 has geometric origin in an Einstein theory on M^{4+N} satisfying the Killing conditions ii and the condition of a bi-invariant isometric immersion of G in M^{4+N}, eq. (3.9).

This fact is sometimes called the Kaluza-Klein "miracle" (A. Salam) or the Kaluza-Klein "isomorphism" (A. Trautman). M^{4+N} is G-compactified in the N vertical directions; the standard (Einstein-Kaluza-Klein) interpretation, of the lack of direct "tactile" evidence for the N extra dimensions, is that the volume of each fibre is too small to be accessible.

4. Generalized Equivalence Principle

Einstein's equivalence principle states that under the influence of pure gravity, mass-points follow worldlines which are geodesics of M^4. In one stroke, this principle summarizes the intrinsic geometric nature of gravitation, thereby justifying its universal character, and identifies the local "normal" coordinate systems, in which particles are locally freely-falling and where the local equivalence between gravitational forces and apparent forces becomes manifest[15].

A guiding idea in the Kaluza-Klein original theory was to formulate a geodesic principle for charged mass points, so as to obtain the charged particle worldlines by projecting onto M^4 the geodesics in M^5. Following this idea, one can directly obtain[9] the Kaluza-Klein theory from the well known equations (on M^4) describing the motion of particles subjected to combined gravitational and Lorentz forces.

A similar procedure can be used, to generate the nonabelian generalization of the Kaluza-Klein: one uses the generalized equivalence principle,[16,17] which states that Yang-Mills charge-masspoints, subject to gravitational and Yang-Mills forces, follow (4+N)-worldlines which are geodesics in $M^{4+N}(M^4,G)$, so as to recover (by projection on M^4) the ordinary worldlines for pointlike Yang-Mills changes. However, in the literature there is some confusion as to what is the correct point-particle limit for Yang-Mills charge-mass distributions. Hence, we[17] proceeded the other way around, i.e., we used the multidimensional theory on M^{4+N} to derive the charge-mass-point equations on M^4. Thus, the generalized equivalence principle now has heuristic value, since it settles the question as to what is the correct point-particle limit for Yang-Mills charges.

For simplicity, I will limit the discussion to a strictly Yang-Mills-type theory, corresponding to the situation (3.9).

As shown in Ref. [17], the geodesic equation on M^{4+N} leads to the equations

$$\frac{d}{dt} u^a + u^b u^d \left\{ {a \atop bd} \right\}^{(4)} + u^b Q^A F_{Ab\cdot}{}^a = 0 \ . \tag{4.1a}$$

$$\frac{d}{dt} Q^A = u^a A_a^B C^A_{\cdot DB} Q^D \ , \tag{4.1b}$$

where (u^a, 0^A) are the components of the geodesic tangent vector in the natural adapted basis (∂_a, ∂_A). u^a defines the four-velocity, while the charges Q^A, having real values, are identified with the ratios e^A/m, where m is the mass of the point particle and $e^A(t)$ its Yang-Mills charge along the group-direction T_A. An equation similar to (4.1a) was written in Ref. [11b]; unfortunately, that paper was plagued by errors. A set of equations resembling (4.1) but with SU(2) matrix generators in place of the numerical charges Q^A, was derived in Ref. [18] from the Dirac-Yang-Mills equations in flat space-time. Similar equations were discussed in Refs. [19], on the basis of an action principle on M^4 but with a rather ad hoc identification of point charges with suitable expectation values of group generators.

Our extremely simple derivation[17] shows the superiority of the geometric approach in M^{4+N} over other approaches. Since the final equations (4.1) contain no dependence on the X-coordinates, group averaging is trivial and (4.1) are directly interpreted as equations for curves on M^4.

It should be noted that the geodesic principle cannot be formulated on M^4, nor on any four-dimensional section in M^{4+N}. Indeed, the section, on which the geodesic $\xi(\tau)$ lies, varies depending on $\hat{\chi}^A(\tau)$. This is one reason why I insist on looking at the theory on M^{4+N} as the fundamental theory.

Causality on M^4 requires that physical particle worldlines be non-spacelike. To guarantee causality on M^4, we must impose causality in M^{4+N}. Since, for a geodesic, "once timelike always timelike", the above property is guaranteed by an initial condition $\gamma(\chi,\chi)|_{\tau_0} \geq 0$. Now, we look at this condition in a normal adapted coordinate system or, even more simply, at some point where $F_{Ab}{}^{a}$ vanishes and g_{ab} is locally Minkowskian. Then we obtain $m^2-e^2 \geq 0$, $e^2 = |e^A e_A|$. In other words, classical point particles with nonvanishing charge cannot be massless.

5. Issue of the Cosmological Constant

The Einstein (4+N)-equations in the lift frame $(\hat{\partial}_a, V_A)$ are simply written as follows:

$$\hat{E}_{ab} = 0 \ , \ \hat{E}_{aA} = 0 \ , \ \hat{E}_{AB} = 0 \ , \tag{5.1}$$

where $\hat{E}_{\alpha\beta}$ are the components of the Einstein tensor.

Specializing to solutions which satisfy the Kaluza-Klein constraint (3.9), eq. (5.1ab) leads directly to the Einstein equations on M^4, with the usual source given by the Yang-Mills energy-momentum density. Similarly, (5.1aA) leads to the covariant conservation law $\nabla \cdot F = 0$. The additional $N(N+1)/2$ vertical equations (5.1AB) have no counterpart in the usual Einstein-Yang-Mills theory on M^4, but in M^{4+N} they are equations which impose further consistency constraints on the theory. If the Kaluza-Klein constraint is abandoned, in favor of the more general Jordan-Thiry theory, then equations (5.1AB) are actually needed in order to have sufficiently many dynamical equations for the vertical metric fields ϕ_{EF}. This is one more instance of the fact that the (4+N)-theory has a richer structure than the dimensionally reduced Einstein-Yang-Mills theory.

Now I return to a point which was glossed over in Sect. 3: the cosmological term $\lambda - R^{(G)}(\{\})$ which occurs in the dimensionally reduced theory when (3.9) is imposed. In order to have flat Minkowski 4-yacuum, so that $R^{(4)} = 0$ for $F^A_{ab} = 0$, we clearly need to set $\lambda = R^{(G)}(\{\})$. On the other hand, in order that the 4-variational problem (3.5-6) have something to do with the (4+N)-problem (3.7), it is clear that we must require that extremals of the 4-theory be group averages of extremals of (3.7). Hence, we must look at the (4+N)-vacuum to see if its group average leads to the flat 4-vacuum.

When $F^A_{ab} = 0$ and the constraint (3.9) holds, the Einstein (4+N)-equations (5.1) considerably simplify, as follows;[3,21]

$$R^{(4)}_{ab}(\{\}) - \frac{1}{2}(R^{(4)}(\{\}) + R^{(G)}(\{\}) - \lambda)g_{ab} = 0, \quad R_{aA} \equiv 0 ,$$
$$R^{(G)}_{AB}(\{\}) - \frac{1}{2}(R^{(4)}(\{\}) + R^{(G)}(\{\}) - \lambda)G_{AB} = 0. \tag{5.2}$$

Contract (5.2ab) with g^{ab} and (5.2AB) with G^{AB} to obtain

$$R^{(4)}(\{\}) = 2(R^{(G)}(\{\}) - \lambda) , \tag{5.3h}$$

$$(\frac{N}{2} - 1)R^{(G)}(\{\}) = \frac{N}{2}(R^{(4)}(\{\}) - \lambda) . \tag{5.3V}$$

If we now impose $\lambda = R^{(G)}(\{\})$, we do get $R^{(4)}(\{\}) = 0$ from (5.3h), but then (5.3V) leaves the unique possibility

$$R^{(G)}(\{\}) = 0 , \tag{5.4?}$$

so that one also needs $\lambda = 0$. Unfortunately, for G nonabelian eq. (5.4?) is wrong. Indeed, for G semi-simple and compact, but otherwise arbitrary, one has $R^{(G)}(\{\}) = N/4$. From this and (5.3) we obtain the unique (4+N)-vacuum solution[21]

$$R^{(4)}(\{\}) = \frac{4R^{(G)}}{N} = 1, \quad \lambda = \frac{1}{2} + \frac{N}{4} . \tag{5.5!}$$

Eq. (5.5) means total disaster: either the group is so "big" that we could be freely walking around in "color" space, or spacetime is so "small" that we could not walk at all.

One can still play with fantastically small constants, to renormalize the relative sizes of $R^{(4)}$ and $R^{(G)}$, but the crucial point remains that the flat Minkowski 4-vacuum is not a solution of the (4+N)-equations, and the (4+N)-theory is in serious trouble.

To avoid this conclusion, e.g. there was a proposal[14] for using peculiar "flat groups" or the idea[22] of adding non-geometric extra terms of θ_{4+N}, thereby spoiling the whole motivation for doing generalized Kaluza-Klein theory.

I took the results (5.5) as an indication that the starting point, eq. (3.7), is in some sense in contrast with the geometry assumed for M^{4+N}. Recall that the fundamental vector fields V_A are left-invariant, so that $V_A|_\xi$ is uniquely determined at all $\xi \in G_x$ --- through the group operations --- from its value at $\xi_o = \sigma_x^{-1}(I)$. Clearly, this is a notion of distant absolute parallelism defined by the group operations. But now, in eq. (3.7) we have introduced a Levi-Civita absolute parallelism, which is dynamically related to the metric through the Einstein equations (5.1). These two connections, for a vertical displacement along V_E of a vertical vector V_A, are as follows: the Levi-Civita symbol gives

$$D_E V_A = \{^{\alpha}_{EA}\}\hat{X}_\alpha = \{^{D}_{EA}\}V_D = \frac{1}{2}\, C^D_{EA} V_D \quad , \tag{5.6}$$

while the group operation gives

$$\nabla_E V_A = [V_E, V_A] = C^D_{\cdot EA} V_D \ . \tag{5.7}$$

Since there is a factor ½-difference, we should motivate our choice of (5.6) rather than the natural group-displacement (5.7). The motivation for choosing the Christoffel connection is only one of analogy with Einstein's general relativity.

However, here we want a fibered structure for M^{4+N}, presumably generated by some peculiar distribution of primordial (4+N)-matter. Thus, it is not at all obvious that the analogy with general relativity on M^4 should be pushed as far as in (3.7).

To decide the issue, I abandon (3.7) and, as in the Palatini approach and in the Einstein-Cartan theory, I consider the metric $\underline{\gamma}$ and a linear connection $\underline{\omega}$ on M^{4+N} as fundamental dynamical variables, related by the Einstein equations:

$$R_{\alpha\beta}(\omega) - \frac{1}{2}\,(R-\lambda)\gamma_{\alpha\beta} = 0 \ . \tag{5.8}$$

I assume that the covariant derivative ∇ defined by $\underline{\omega}$ satisfies the metricity condition: $\nabla\underline{\gamma} = 0$, which fixes the symmetric part of $\underline{\omega}$: $\omega^\alpha_{(\beta\gamma)} = \{^\alpha_{\beta\gamma}\}$. To determine torsion, I require that, since spin-½ fields are absent in the reduced theory on M^4, all components of torsion along horizontal directions vanish. The vertical component $\omega^A_{\ BC}$ of torsion is still undetermined; since it has vanishing shadow on M^4, we need some new recipe to determine it. This recipe is found as follows:

Let $\xi \longrightarrow \xi\cdot Y$, Y an infinitesimal element of G, be the motion on M^{4+N} generated by $Y^F V_F$. Let (ω,γ) be a solution of (5.11), and $\hat{\omega}^\alpha_\beta = \hat{\omega}^\alpha_{\delta\beta}\hat{\kappa}^{*\delta}$ the associated connection 1-forms in the coframe $\hat{\kappa}^{*\delta}$ dual to $\hat{\kappa}_\delta$. I define the gauge transformations

$$\delta\hat{\kappa}_\alpha \equiv Y^F[V_F,\hat{\kappa}_\alpha], \text{ i.e. } \delta V_A = Y^F C^D_{\cdot FA} V_D, \quad \delta\hat{\omega}^\alpha_\beta \equiv Y^F \underset{V_F}{L}\, \hat{\omega}^\alpha_\beta ,$$

$$\text{i.e. } \delta\hat{\omega}^A_{DB} = -Y^F C^E_{\cdot FA}\hat{\omega}^D_{EB} . \tag{5.9}$$

I require that $(\omega + \delta\omega, \gamma + \delta\gamma)$ also be a solution of (5.8), so that these gauge transformations are symmetries of the theory (5.8). It turns out that, when (3.9) is imposed, only two connections ω^D_{AB} are allowed by this requirement of "strong" gauge invariance:

$$\hat{\omega}^D_{AB} = C^D_{\cdot AB} \quad \text{or} \quad \hat{\omega}^D_{AB} = 0 . \tag{5.10}$$

For reasons previously explained, I prefer the first choice, but the second choice is as good. Indeed, one easily shows that, for the connections (5.10), one has

$$R^{(G)}_{AB}(\hat{\omega}^D_{AB}) \equiv 0 . \tag{5.11}$$

Now we can repeat the little game, which led from (5.2) to (5.4), except that now we have (5.11). It is easily seen that there is again a unique vacuum solution determined by (5.11):

$$R^{(4)}_{ab} = 0 = R^{(G)}_{AB} , \quad \lambda = 0 . \tag{5.12}$$

This is a prediction, of a flat 4-vacuum ($R^{(4)} = 0$, $\lambda = 0$) from dimensional reduction of a G-symmetric (4+N)-vacuum.

The new vacuum connection, $\hat{\omega}^D_{AB} = C^D_{AB}$ (or zero), has torsion, so that the theory is not completely "metric". However, this torsion is purely vertical: it has vanishing effect on M^4, except for the implication $\lambda = 0$. Nonmetric extensions of the Kaluza-Klein nonabelian theory have also been considered in Ref. [23] (I thank Prof. A. Trautman for calling this work to my attention). There, a connection was used, which had horizontal and vertical torsion components; with that connection, it was impossible to satisfy the metricity condition $\nabla\underline{\gamma} = 0$, and the dimensionally reduced theory was not the same as the Einstein-Yang-Mills theory. These problems are not present in our approach.

References

1. A. Trautman, Rep. Math. Phys. (Terun) 1 (1970), 29; also in Lectures on General Relativity, ed. by S. Deser and K.W. Ford, Prentice-Hall 1965, and in General Relativity and Gravitation, ed. by A.H. Held, Plenum Press, 1980.
2. S. Kobayashi and K. Nomizu, Foundations of Differential Geometry, Interscience 1963.
3. C.A. Orzalesi, Fortschr. d. Physik 29 (1981) 433.
4. Y.M. Cho, J. Math, Phys. 16 (1975), 2029;
   Y.M. Cho and P.S. Chang, Phys. Rev. D12 (1975), 3789.
5. M. Toller, Nuovo Cimento B44 (1978), 67;
   G. Cognola, R. Soldati, L. Vanzo and S. Zerbini, J. Math. Phys. 20 (1979) 2613.
6. T. Kaluza, Sitzb. Preuss. Akad. Wiss. (1921), 966;
   O. Klein, Z. Phys. 37 (1926) 895.
7. A. Einstein, Sitzb. Preuss. Akad. Wiss. (1927), 23;
   A. Einstein and P. Bergmann, Ann. Math. 39 (1938), 683.
8. P. Jordan, Astr. Nachr. 276 (1948), 193;
   Y.R. Thiry, Compt. Rend. 226 (1948), 216.
9. A. Lichneronicz, Théories Relativistes de la Gravitation et de l''Electromagnetisme, Masson Paris 1955.
10. B. De Witt, in Relativity, Groups and Topology, ed. by B. De Witt and C. De Witt, Gordon and Breach (1964).
11. (a) J. Rayski, A. Phys. Pol. 27 (1965) 89; (b) R. Kerner, Ann. Inst. H. Poincaré 9 (1968), 143.
12. Y.M. Cho and P.G.O Freund, Phys. Rev. D12 (1975), 1711.
13. L.N. Chang, K.I. Macrae and F. Mansouri, Phys. Rev. D13 (1976) 235.
14. J. Scherk and J.H. Schwartz, Nucl. Phys. B153 (1979) 61.
15. C.W. Misner, K.S. Thorne and J.A. Wheeler, Gravitation, Freeman & Co. (1973).
16. C.A. Orzalesi, in Particle Physics 1980, ed. by I. Andrić, I. Dadić and N. Zovko, North-Holland 1981.
17. C.A. Orzalesi and M. Pauri, Nuovo Cimento, to be published.
18. S.K. Wong, Nuovo Cimento 45A (1970) 689.
19. A.P. Balachandran, P. Salomonson, B.S. Skagerstam and J.O. Winnberg, Phys. Rev. D15 (1977) 2308;
    P. Salomonson, B.S. Skagerstam and J.O. Winnberg, ibid 16 (1977) 2581.
20. W. Pauli, Teoria della Relatività, Boringhieri (1958).
21. C.A. Orzalesi and M. Pauri, Physics Letters B, to be published.
22. F. Luciani, Nucl. Phys. B135 (1978) 111.
23. W. Kopczynski, Acta Phys. Pol. B10 (1979) 365.

ON THE SYMMETRY PROPERTIES OF SEPARATED MONOPOLE CONFIGURATIONS

by

L. O'Raifeartaigh and S. Rouhani

Dublin Institute of Advanced Studies,
10 Burlington Road, Dublin 4, Ireland.

ABSTRACT: It has been known for some time that when the potential is zero and the Higgs field is in the adjoint representation, the Yang-Mills-Higgs system admits regular solutions describing any number of magnetic monopoles in static equilibrium. The explicit construction of such solutions for SU(2) by means of CP_3 transition matrix is described, and the symmetry properties (notably axial symmetry properties) are investigated. It is shown that solutions corresponding to rings of monopoles with discrete axial symmetry are permitted, and that for small monopole separations the general solution may be expressed as a linear superposition of such ring solutions.

1. Introduction

The discovery in 1974 that non-Abelian Yang-Mills-Higgs (YMH) systems admit magnetic monopoles of finite (non-zero) extension and finite (non-infinite) mass[1)] has led to an extensive study of such objects. The first such studies concentrated on the classification of the monopoles according to their topology[2)] but in recent times the interest has shifted to the actual construction of monopole solutions to the YMH equations. The interest has focused primarily on the special case in which the Higgs potential V vanishes (after spontaneous breakdown) and the Higgs field Φ belongs to the adjoint representation of the YM group[3)]. Although this special case may be unphysical (because $V=0$ implies massless Higgs fields) it has attracted interest for two reasons. First, when $V=0$ and $\Phi\varepsilon$ adjoint the YMH system admits not only single monopoles of arbitrary strength but also systems of separated monopoles in static equilibrium[4)]. Second when $V=0$ and $\Phi\varepsilon$ adjoint, the static YMH equations reduce to a system of self-dual equations, analogous to those which proved so tractable for instantons[5)]. At the time of the 1980 Clausthal conference the existence of static separated monopole configurations had been established, but neither these solutions, nor single-monopole solutions of strength greater than unity, had been constructed. Indeed it was known that a direct attack on the problem of constructing solutions, using symmetry principles, would not be easy, because, for SU(2), it had been shown that single-monopole solutions of strength greater than unity could not be spherically symmetric[6], and the two separated monopole solutions could not be axially symmetric[7]. (In fact, the latter result was the content of my talk at Clausthal last year.)

During the year that has elapsed since the Clausthal conference, explicit solutions of the SU(2) YMH equations for $V=0$ and $\Phi\varepsilon$ adjoint have been constructed[8-12]. These solutions describe both single monopoles of arbitrary strength and separated monopoles, and they may well include all the solutions because they contain, in general, the number of parameters predicted by the index theorem[13]. The purpose of the present talk is to describe how the explicit solutions have been constructed, and to consider their symmetry properties. It turns out that for monopole strength greater than unity the energy-distribution is not spherically symmetric, and is not axisymmetric unless the monopoles are superposed[12], and this result goes some way toward explaining the negative results on symmetry mentioned earlier. Furthermore, for separated monopoles, we shall show here that solutions corresponding to rings of monopoles with discrete axial symmetry (of order equal to the number of monopoles on the ring) are permitted, and this result also helps to explain the earlier result, since it implies that a continuous ring of monopoles (and hence an infinite mass) would be required to have continuous axial symmetry. We shall also show that for small monopole separations at least, the general solution can (at a certain level) be expressed as a linear superposition of the discrete axisymmetric ring solutions. Finally, we should mention that the twisted axial symmetry of the two-monopole system which I discussed during my talk is not correct. However, the discrete symmetries and monopole locations which I discussed are correct, and, in fact, the rings of monopoles described in Section 4 are the generalization of these results to n monopoles.

2. Construction of Solutions

There are actually three approaches to the problem of finding the solutions to the YMH-system for $V=0$, $\Phi\varepsilon$ adjoint now in existence, namely the inverse scattering method[12], the instanton method[10] and the transition matrix method due to Ward[8], but we shall consider only the latter method (which was also the first). The first step for any method is to note that for $V=0$ and $\Phi\varepsilon$ adjoint, the static YMH Hamiltonian can be written in the form[14]

$$\mathcal{H} = \frac{1}{2}\int d^3x[\vec{B}^2 + (\vec{D}\Phi)^2] = Q + \frac{1}{2}\int d^3x(\vec{B} - \vec{D}\Phi)^2 \quad , \tag{2.1}$$

where $\vec{B}$ is the gauge-field ($\vec{E}=0$), $\vec{D}$ the covariant derivative, and Q is the topological charge. It is then evident that the equations which minimize the energy (at the value Q) are the so-called Bogomolny equations

$$\vec{B} = \vec{D}\Phi \ . \tag{2.2}$$

The boundary conditions induced by the potential before setting $V=0$ are $\vec{B} \to 0$ and $\Phi^2 \to 1$ as $|\vec{x}| \to \infty$. Note that the equations (2.2) are first-order. The next step is to note that if we let $F_{\mu\nu}$ be a 4-dimensional Euclidean gauge-field, satisfying the self-dual equations

$$F_{i4} = \frac{1}{2} \varepsilon_{ijk} F_{jk} \ , \qquad i,j,k = 1,2,3 \tag{2.3}$$

then, if $F_{\mu\nu}$ is independent of x_4 and Φ is identified with A_4, Eqns. (2.2) and (2.3) become identical. Thus the Bogomolny equations may be interpreted as the special (x_4-independent) case of self-dual equations.

The Ward approach is now based on the fact that every self-dual field (2.3) on E_4 is equivalent to an SU(2) transition matrix on the 3-dimensional complex projective 3-space CP_3. More precisely if μ,ν,ξ are (inhomogeneous) complex coordinates for CP_3 and $\mu = t+z+x_-\xi^{-1}$, $\nu = t-z+x_+\xi$, where $x_\pm = x \pm iy$, then any static solution of (2.3) is equivalent[15] to an SU(2) matrix of the form

$$g(\gamma,\xi) = \begin{bmatrix} (e^{\kappa}+(-1)^n e^{-\kappa})/H & (-\xi)^{-n} e^{-\kappa} \\ (\xi)^n e^{-\kappa} & He^{-\kappa} \end{bmatrix} , \tag{2.4}$$

where $\gamma = \mu-\nu$, H is a polynomial in γ,ξ and ξ^{-1} of degree n in ξ and ξ^{-1}, and K is a polynomial of degree $n-1$ in γ which is related to H as discussed below. The explicit method of obtaining the $F_{\mu\nu}(x)$ from $g(\gamma,\xi)$ is given in Ref. 16 and consists essentially in carrying out some integrations with respect to θ where $\xi = e^{i\theta}$ and then making some non-linear algebraic manipulations.

Since the Eqns. (2.2) and (2.3) are automatically satisfied by (2.4) the problem is reduced to choosing the functions H,K so as to satisfy the boundary and regularity conditions for $\vec{B}$ and Φ. This problem was solved by Ward[8], first for two superimposed monopoles and later for two separated monopoles. Both cases were later generalized to the case of n monopoles[9-12] and the present Ansatz for n separated monopoles is

$$H = \gamma^n + a_1(\xi)\gamma^{n-1} + \ldots a_n(\xi) = \prod_{r=1}^{n} (\gamma - \gamma_r(\xi)) \quad , \tag{2.5}$$

where the $a_k(\xi)$ are polynomials of degree k in ξ and ξ^{-1}, and

$$K = i\pi \sum_{k=1}^{n} n_k \prod_{\ell \neq k} \left(\frac{\gamma - \gamma\ell}{\gamma k - \gamma\ell} \right) = b(\xi)+b_o(\xi)\gamma+b_1(\xi)\gamma^2+\ldots+ b_{n-2}(\xi)\gamma^{n-1}, \tag{2.6}$$

where the n_k are the first n even or odd (positive and negative) integers, and where K must satisfy the $n(n-2)$ Corrigan-Goddard (CG) conditions

$$\frac{1}{2\pi i} \oint_{|J|=1} \frac{d\xi}{\xi} b_s \xi^m = 0 \quad , \qquad \begin{matrix} -s \leqslant m \leqslant s \quad , \\ s = 1\ldots n-2 \quad . \end{matrix} \tag{2.7}$$

Note that the function K in (2.7) is completely determined by H. We shall therefore call H the generating function for the solutions. Note that it contains $n(n+2)$ parameters, but since there are $n(n-2)$ conditions in (2.7) there are only $4n$ independent parameters altogether ($4n-1$ if the norm of the Higgs field is fixed as in (2.2)). Since the solutions are completely determined by the generating function H, so are their symmetries, and hence to consider the symmetries, as we shall now do, it suffices to consider the symmetry properties of H.

3. The Euclidean Group

Let us first consider the properties of H with respect to the Euclidean group. For this purpose it is convenient to replace (μ,ν,ξ) by the homogeneous coordinates (μ,ν,π_1,π_2), where

$$\begin{pmatrix} \lambda_1 \\ \lambda_2 \end{pmatrix} = \begin{pmatrix} t+z & x_+ \\ x_- & t-z \end{pmatrix} \begin{pmatrix} \pi_1 \\ \pi_2 \end{pmatrix} , \qquad \begin{aligned} \mu &= \lambda_1/\pi_1 \\ \nu &= \lambda_2/\pi_2 \\ \xi &= \pi_2/\pi_1 \quad , \end{aligned} \tag{3.1}$$

and introduce the quantities $\vec{\Sigma} = i\tilde{\pi}\sigma_2\vec{\sigma}\pi$ where π is the column-vector (π_1,π_2), tilde denotes transpose and $\vec{\sigma}$ are the Pauli matrices. One sees from (3.1) that $\vec{\Sigma}$ is a translational invariant and a rotational vector. Once also sees that $\Sigma^2 = 0$ and hence the definition of $\vec{\Sigma}$ can be inverted to give $\xi = \Sigma_+/\Sigma_3$ and $\xi^{-1} = -\Sigma_-/\Sigma_3$. From (2.5) one then sees that H may be written in the form

$$H\Sigma_3^n = w^n + w^{n-1}(\vec{a}\cdot\vec{\Sigma}) + w^2(a_{ij}\Sigma_i\Sigma_j) + \ldots + (a_{i_1\ldots i_n}\Sigma_{i_1}\ldots\Sigma_{i_n}) \quad , \tag{3.2}$$

where $\omega = \gamma\Sigma_3 = (\vec{x}-\vec{\Sigma})$. The factor Σ_3^n following H may actually be gauged to unity. Equation (3.2) exhibits the Euclidean properties of H explicitly. In particular we see that w is rotationally invariant but transforms under translations $\vec{x} \to \vec{x} + \vec{x}_0$ according to $w \to w + \vec{x}_0 \cdot \vec{\Sigma}$, while the $n(n+2)$ parameters $a_{i_1\ldots i_r}$ may be regarded as (translationally invariant) components of real traceless symmetric of order $r \leqslant n$. Thus the parameter space consists of SO(3) orbits.

One use of the Euclidean group is to reduce the number of parameters by six. For example, for $n=2$ the term linear in ω can be translated to zero and the term without ω rotated so that

$$H = (\omega^2 + a_0\Sigma_3^2 + a_1\Sigma_1\Sigma_3)/\Sigma_3^2 = \gamma^2 + a_0 + a_1(\xi - \xi^{-1}) \quad , \tag{3.3}$$

or, alternatively,

$$H = [\omega^2 + a_0\Sigma_3^2 + a_2(\Sigma_+^2 + \Sigma_-^2)]/\Sigma_3^2 = \gamma^2 + a_0 + a_2(\xi^2 + \xi^{-2}) \quad . \tag{3.4}$$

The choice (3.3) is the one originally made by Ward[8], but (3.4) has the advantage that the solution is invariant under reflexions in the coordinate planes.

4. Symmetries of H and Rings of Monopoles

It is evident from (3.2) that H will be rotationally invariant only if all the coefficients $a_{i_1\ldots i_r}$ are zero. But then $H=\gamma^n$ and this is known[11] to lead to a singular solution unless $n=1$. This is why the spherically symmetric SU(2) monopoles must have charge unity. Let us consider a more restricted set of invariance groups of H , namely, those which may be implemented by linear transformations on ξ , or more precisely, linear transformations of $\vec{x}$ and $\vec{\xi}$ which leave $\gamma = 2z + x_+\xi - x_-\xi^{-1}$ invariant (or change its sign). This set of groups includes reflexions in the coordinate planes and axial rotations, both discrete and continuous.

The discrete symmetries have been tabulated in Ref. 17 and hence we concentrate on axial symmetries, for which $x_+ \to x_+e^{i\alpha}$, $\xi \to \xi e^{-i\alpha}$.

In the continuous case, $0 \leq \alpha < 2\pi$, it is evident that H will be symmetric only if all the coefficients are zero except $a_{o\ldots o}$. The CG conditions and regularity then fix the $a_{o\ldots o}$ to be such that $H \prod_{r=1}^{n}(\gamma - n_r)$ where n_r are the integers occurring in (2.6) and this H is known to describe n superimposed monopoles, in agreement with last year's result on axial symmetry.

The interesting point, however, is that if we relax the continuous condition $0 \leq \alpha < 2\pi$ and allow α to take only certain discrete values then it is possible to obtain separated monopole solutions. In particular let us consider the case when α takes the discrete values $\alpha = 2\pi r/n$, $r = 1\ldots n$, and thus generates the discrete axial group C_n . Then H is necessarily of the form

$$H = \gamma^n + a_1\gamma^{n-1} + \ldots + a_n + \varepsilon\xi^n + \bar{\varepsilon}(-\xi)^{-n} \tag{4.1}$$

where the a_r and ε are constant and the a_r are real. In that case the function K will depend only on ξ^n and ξ^{-n} , so all the CG conditions will be automatically satisfied, except for the n-2 conditions

$$\frac{1}{2\pi i} \oint \frac{d\xi}{\xi} b_s(\xi^n) = 0 \quad , \tag{4.2}$$

and these simply normalize n-2 of the a_r's . Actually, if we impose also z- and y- reflexion invariance (4.1) and (4.2) reduce further[17] to

$$H = \gamma^n + a_2\gamma^{n-2} + \ldots + a_{2n-2m}\gamma^{n-2m} + \varepsilon(\xi^n + (-\xi)^{-n}) \ , \text{ and } \frac{1}{2\pi i}\oint \frac{d\xi}{\xi} b_{2s} = 0 \ , \tag{4.3}$$

respectively, where n=2m or 2m+1 .

To find out what kind of monopole configuration is described by (4.1)-(4.3) we note that the discrete axial symmetry implies that it must describe either a ring of n monopoles or a set of n monopoles on the z- axis. In the latter case, however, the Higgs field would have to vanish on the axis, and a direct computation shows that it does not. Thus (4.1)-(4.3) describe rings of monopoles. It should, perhaps, be mentioned that for n=2 and n=3 the CG conditions in (4.3) are automatically satisfied. The case n=2 is the original separated solution of Ward (in the symmetrical coordinate system (3.4) while n=3 is a new explicit solution describing 3 equidistant monopoles. For all n the only free parameters are the norm of the Higgs field and the radius of the ring.

More generally, if we choose

$$H = \gamma^n+a_1\gamma^{n-1}+\ldots+a_{\ell-1}\gamma^{n-\ell+1}+[\varepsilon\xi^\ell+\bar{\varepsilon}(-\xi)^{-\ell}](\gamma^{n-\ell}+a_{\ell+1}\gamma^{n-\ell-1}+\ldots+a_n) \ , \tag{4.4}$$

then the system will be invariant with respect to the discrete axial symmetry group C_ℓ , or $\alpha = 2\pi r/\ell$, $r = 1\ldots\ell$, and the CG conditions will reduce to

$$\frac{1}{2\pi i}\oint\frac{d\xi}{\xi}\, b_s(\xi^\ell)(\xi^\ell)^r = 0 \ , \qquad \text{for} \quad r\ell \leqslant n-2 \ . \tag{4.5}$$

Such a system will presumably describe a ring (or rings) of monopoles (with some monopoles on the z- axis if n/ℓ is not integral) though for $\ell<n$ we have not checked this explicitly that all the monopoles do not lie on the axis.

5. General Solution for Small Separations: Superpositions of Rings

If no symmetry is imposed on the generating function H there seems to be no way in which the $4n$ independent parameters can be extracted from the CG conditions. Indeed, in general, the CG conditions seem to be transcendental. However, the $4n$ independent parameters can be found explicitly in the case that all the parameters are close to their axisymmetric (superimposed monopole) values, that is, for

$$H = \prod_{k=1}^{n}(\gamma-n_k) + \sum_{k=1}^{n}\sum_{m=o}^{k}[\varepsilon_k^{(m)}\xi^m + \bar{\varepsilon}_k^{(m)}(-\xi)^{-m}]\gamma^k \ , \tag{5.1}$$

where the n_k are the integers occurring in (2.6) and the $\varepsilon_k^{(m)}$ are small. It turns out[18] that to first order in $\varepsilon_k^{(m)}$ the CG conditions (2.7) reduce to a set of $n(n-2)$ <u>algebraic</u> conditions of the form

$$\sum_{k=1}^{n}\rho_{\alpha k}\varepsilon_k^{(m)} = 0 \ , \qquad \alpha = 1\ldots n-m-1 \ (m\neq o) \ , \quad \alpha = 1\ldots n-2 \ (m=o) \ , \tag{5.2}$$

where ρ_{ik} is the numerical matrix

$$\rho_{ik} = \sum_{r=1}^{n}\frac{(n_r)^{i+k-2}}{[\prod_{s\neq r}(n_s-n_r)]^2} \ , \qquad i,k = 1\ldots n-m+1 \ (m\neq o) \ , \ i,k=1\ldots n \ (m=o). \tag{5.3}$$

The equations (5.2),(5.3) can be solved immediately to give

$$\varepsilon_k^{(m)} = (\rho^{-1})_{k\lambda}\varepsilon_\lambda^{(m)} \;, \tag{5.4}$$

where λ takes only the two end-values $\lambda=n-m\,,\,n-m+1$ for the generic case $1\leqslant m\leqslant n-1$ and the end-values $\lambda=n-1,n$ and $\lambda=1$ for the special cases $m=0$ and $m=n$ respectively. Eq. (5.4) expresses the $n(n+2)$ parameters $\varepsilon_k^{(m)}$ as linear combinations of the $4n$ parameters $\varepsilon_k^{(m)}$. Returning to (5.1) we see that

$$H = \prod_{k=1}^{n}(\gamma-n_k) + \sum_{k=1}^{n}\sum_{m=o}^{k}(\rho^{-1})_{k\lambda}[\xi_\lambda^{(m)}\xi^m + \bar{\varepsilon}_\lambda^{(m)}(-\xi)^{-m}]\gamma^k \;, \tag{5.5}$$

gives the complete solution to the monopole problem for small $\varepsilon_\lambda^{(m)}$ (and hence for small separations).

Note that only the dimension of the matrix ρ and not the values of its elements, depend on m, and thus there are only two independent parameters for each m in (5.5). Note also that ρ has no elements connecting even and odd values of i,k and hence that (5.4) may be decomposed into two parts

$$\varepsilon_k^{(m,+)} = (\rho^{-1})_{k\tau}\,\varepsilon_+^{(m)} \quad \text{and} \quad \varepsilon_k^{(m,-)} = (\rho^{-1})_{k\tau}\,\varepsilon_-^{(m)} \;, \tag{5.6}$$

where τ takes only a <u>single</u> end-value in each case. If one observes that the values of m correspond to irreducible representations of the axial rotation group C_∞ and the $\pm$ in (5.6) to parity, one sees that the $4n$ parameters have a simple group-theoretical interpretation: they are the excitations corresponding to the lowest $4n$ irreducible representations of the group C_∞x Parity.

The generating function (5.5) has also a simple interpretation in terms of the rings of monopoles discussed in section 4. Indeed, if all the parameters $\varepsilon_\pm^{(m)}$ are chosen to be zero except one, $\varepsilon_+^{(m_o)}$ say, the function (5.5) reduces to the ring solution (4.4). Thus the general function (5.5) (for small $\varepsilon_\pm^{(m)}$) may be regarded as a linear superposition of $4n$ ring functions. Of course, the superposition is only at the level of the generating function H, and the Higgs field, and its zeros, will not be linear superpositions.

References

1. G.'t Hooft, Nucl. Phys. B79, 276 (1974); A. Polyakov, JETP Lett. 20, 194 (1974)
2. P. Goddard, D. Olive, Rep. on Prog. Phys. 41, 1357 (1978)

3. M. Prasad, C. Sommerfield, Phys. Rev. Lett. 35, 760 (1975)
4. C. Taubes, Commun. Math. Phys. 80, 343 (1981)
5. M. Atiyah, N. Hitchin, V. Drinfeld, Y. Mannin, Phys. Lett. 65A, 185 (1978)
6. A. Guth, Weinberg, Phys. Rev. D14, 1660 (1976); L. O'Raifeartaigh, Nuovo Cim. Lett. 18, 205 (1976)
7. P. Houston, L. O'Raifeartaigh, Proc. Conf. Diff. Geom. Methods, Clausthal, 1980 (ed. Doebner, Springer, Berlin 1981)
8. R. Ward, Commun. Math. Phys. 79, 317 (1981)
9. E. Corrigan, P. Goddard, Commun. Math. Phys. 80, 575 (1981)
10. W. Nahm, CERN - Preprint (1981)
11. M. Prasad, Commun. Math. Phys. 80, 137 (1981), L. O'Raifeartaigh, S. Rouhani, Schladming (1981) Lectures, Acta Physica Austriaca
12. Z. Horvath, P. Forgacz, L. Palla, Hung. Acad. Sci. Inst. Phys. Preprints KFKI - 21,23 (1981)
13. E. Weinberg, Phys. Rev. D20, 936 (1979)
14. E. Bogomolny, Sov. J. Nucl. Phys. 24, 449 (1976)
15. N. Hitchin, Oxford Univ. Math. Inst. Preprint (1981)
16. E. Corrigan, D. Fairlie, P. Goddard, R. Yates, Commun. Math. Phys. 58, 223 (1978)
17. L. O'Raifeartaigh, S. Rouhani, DIAS Preprint, STP-81-31
18. L. O'Raifeartaigh, S. Rouhani, L.P. Singh, DIAS Preprint, STP-81-33.

HOLONOMY GROUPS IN GRAVITY AND GAUGE FIELDS

Jeeva Anandan
Department of Mathematics
University of California, Berkeley
Berkeley, California 94720

I. Introduction.

It has been pointed out that the phase shift in quantum interference due to the gravitational field or gauge field connection is determined by the holonomy transformations associated with this connection [1]. Now, the quantum mechanical motion of a particle may be regarded as being modified by the external field, via Huygen's principle, by the phase shift in the interference of secondary wave-lets. This suggests that, from a physical point of view, holonomy transformations contain all the observable information in a connection.

In section II it will be shown that, from a mathematical point of view also, holonomy transformations contain all the gauge invariant information in a connection. Indeed, an explicit rule will be given for constructing the connection, which is uniquely determined up to gauge transformations, from the holonomy transformations associated with closed curves that begin and end at <u>any</u> fixed point. When the holonomy group is a subgroup of certain matrix groups, the Wilson loops (traces of holonomy transformations) are sufficient to construct the connection. In section III the concept of holonomy map is used to provide a unified description of gravity and gauge fields. It is then pointed out, in section IV, that Newtonian gravity may be regarded as a gauge field of the group of Galilei boosts.

II. Holonomy Maps.

Let Γ be a connection in a principal fiber bundle over a base manifold M with structure group G and projection map π. Let $x \in M$ and L_x denote the set of all piece-wise differentiable closed curves (loops) at x. For each $\ell \in L_x$, the parallel displacement along ℓ determined by Γ is called the holonomy transformation associated with ℓ and will be denoted by $\underline{\ell}$. Then $\underline{\ell}$ is an automorphism of the fiber F_x over x, i.e. a differentiable transformation of F_x which commutes with the right action of G. Clearly, the set of all such automorphisms form a group, called the holonomy group H_x. It is also easy to show that H_x and $H_{x'}$ are isomorphic for any $x, x' \in M$, which can be joined by a curve.

If $\ell, \ell' \in L_x$ such that their images differ by the images of loops each of which encloses no area then clearly the holonomy transformations associated with ℓ and ℓ' are the same. It is therefore natural to identify such elements of L_x and the resulting set will be denoted by $\mathcal{L}_x$. Since two loops can be composed to give a third loop, this defines a product on $\mathcal{L}_x$. Clearly $\mathcal{L}_x$ is a group under this product and will be called the loop group at x. It follows that the connection Γ determines a homomorphism from $\mathcal{L}_x$ into H_x defined by the assignment to each loop ℓ at x the corresponding holonomy transformation $\underline{\ell}$. This map, with H_x regarded as an abstract group, will be called the <u>holonomy map of Γ</u> at x. Let $\phi : L_x \to G$ be a differentiable map such that it has the same value for any two loops that are identified as mentioned above. Then ϕ may be regarded as a map from $\mathcal{L}_x$ into G. Let $\mathcal{F}_x$ denote the set of such maps. We now prove the following theorem which was first stated by Kobayashi [2]:

Theorem 1: Let M be a manifold and $\phi \in \mathcal{F}_x$ is a homomorphism from $\mathcal{L}_x$ $(x \in M)$ into an arbitrary Lie group G. Then there exists a principal fiber bundle P over M with structure group G and connection Γ such that ϕ is the holonomy map of Γ at x.

Proof: Consider the set of piece-wise differentiable curves in M which originate at x. Two such curves will be called equivalent when their end-points are the same and they enclose no area. Each equivalence class will be called a <u>path</u> originating at x and the set of such paths will be denoted by C_x. Clearly, given $p, p' \in C_x$ with the same end-point, $p' = pc$ for some $c \in \mathcal{L}_x \subset C_x$ where the product pc is defined in the obvious way. Define $\eta : C_x \to M$ as the map which assigns to each path its end point. Given the homomorphism $\phi : \mathcal{L}_x \to G$, $\mathcal{L}_x$ acts differentiably on $C_x \times G$ according to the rule $(p, g) \to (pc, \phi(c)^{-1}g)$, $c \in \mathcal{L}_x$. Let $P = C_x \times G/\mathcal{L}_x$. We shall denote the element of P which contains (p, g), $p \in C_x$, $g \in G$ by $\overline{(p, g)}$. Define $\pi : P \to M$ by the rule $\overline{(p, g)} \to \eta(p)$. Also let G act on P to the right according to $\overline{(p, g)} \to \overline{(p, g)}h = \overline{(p, gh)}$, $h \in G$. Given any $y \in M$, we shall suppose that U is a neighborhood of y such that the points in U are connected to x by a set of curves which vary continuously and differentiably. Let K_u denote the set of paths corresponding to these curves. Then the map $f_u : \Pi^{-1}(U) \to U \times G$ defined by $\overline{(p, g)} = \overline{(p_0, \phi(c)^{-1}g)} \to (\eta(p_0), \phi(c)^{-1}g)$, where $p_0 = pc \in K_u$, obviously commutes with the right action of G. Using maps of the form f_u, P can be given a differentiable structure. It follows also that P is locally trivial and it is a principal fiber bundle over M with G as the structure group and Π as the projection map.

Suppose q is a path joining $y, z \in M$ and p is a path which joins x to y. Define $\tau_q: \Pi^{-1}(y) \to \Pi^{-1}(z)$ by $\overline{(p, g)} \xrightarrow{\tau_q} \overline{(qp, g)}$ for every $g \in G$. Suppose now that $y, z \in U$, $p_0, p_0' \in K_u$, $\eta(p_0) = y$, $\eta(p_0') = z$. Then $\overline{(p_0, g)} \xrightarrow{\tau_q} \overline{(qp_0, g)} = \overline{(p_0', \phi(p_0'^{-1}qp_0)g)}$. Since $\phi \in \mathcal{F}$, the lift of q by τ_q is differentiable. Also τ_q commutes with the right action. Hence τ_q is the parallel displacement with respect to a connection Γ. Let o be the identity of $\mathcal{L}_x$. Then $\overline{(o, g)} \in \Pi^{-1}(x)$. Also, given $\ell \in \mathcal{L}_x$, $\overline{(o, g)} \xrightarrow{\tau_\ell} \overline{(\ell, g)} = \overline{(o, \phi(\ell)g)} \in \Pi^{-1}(x)$. Hence ϕ is the holonomy map of Γ at x.

The question now arises as to how unique are the bundle and connection determined by ϕ. This is answered by the following theorem which is a special case of a theorem proved earlier [3]:

Theorem 2: Let P(M, G) and $\tilde{P}(M, G)$ be two principal fiber bundles over M with projection maps Π and $\tilde{\Pi}$ and right actions $R_a (a \in G)$ and $\tilde{R}_b (b \in G)$ and connections Γ and $\tilde{\Gamma}$ respectively. Then there exists an isomorphism $f: P(M, G) \to \tilde{P}(M, G)$ which maps Γ into $\tilde{\Gamma}$ and whose induced map on M is the identity if and only if for any $x \in M$ there exists a differentiable bijection $g: \Pi^{-1}(x) \to \tilde{\Pi}^{-1}(x)$ such that (i) $gR_a = \tilde{R}_a g$ for every $a \in G$ and (ii) for every loop γ at x, the holonomy transformations $\underline{\gamma}$ and $\underline{\tilde{\gamma}}$ in P(M, G) and $\tilde{P}(M, G)$ are related by $\underline{\gamma} = g^{-1}\underline{\tilde{\gamma}}g$.

It follows from theorems 1 and 2 that the homomorphism $\phi : \mathcal{L}_x \to G$ determines a connection, whose holonomy map at x is ϕ, uniquely up to gauge transformations. If G is regarded as a subgroup of a matrix group by means of an isomorphism then clearly ϕ may also be regarded as a representation. The character of ϕ (i.e. traces of elements in the image of ϕ) is then called the Wilson loop functional. When the image of ϕ is a compact group or a finite dimensional semi-simple group it is known that the character of ϕ determines ϕ up to conjucacy. It follows then that under these circumstances, the Wilson loop functional determines the connection up to gauge transformations.

But $\mathcal{L}_x$ is an infinite dimensional space. Does there exist a finite dimensional manifold $\mathcal{S}_x \subset \mathcal{L}_x$ such that the specification of ϕ on $\mathcal{S}_x$ is sufficient to determine the connection up to gauge transformations?* To answer this question, suppose that M is an n-dimensional manifold which admits a Riemannian or psuedo-

* This question was first asked by I.M. Singer during a seminar.

Riemannian metric such that for any $y, z \in M$, there exists a unique geodesic through y, z. Then for any pair $y, z \in M$, there exists a unique triangular path of geodesics from x to y, y to z and z to x. Let $\mathcal{S}_x \subset \mathcal{L}_x$ be the set of such triangular paths which have x as a fixed vertex. Then $\phi|_{\mathcal{S}_x}$ can also be regarded as a map $\tilde{\phi}$ from the 2n-dimensional manifold $M \times M$ into G satisfying the conditions a) $\tilde{\phi}(y, y) = $ identity, for every $y \in M$ and b) if $y, z, u \in M$ lie on a geodesic then $\tilde{\phi}(y, z)\tilde{\phi}(z, u) = \tilde{\phi}(y, u)$. If $\tilde{\phi}$ is given satisfying (a) and (b) then ϕ can be determined as follows. Let $c \in \mathcal{L}_x$ and let $x, y_1, y_2, \ldots$ divide c into infinitesimal geodesics. Then $\phi(c) = \tilde{\phi}(x, y_1)\tilde{\phi}(y_1, y_2)\tilde{\phi}(y_2, y_3)\ldots$ We therefore have:

Theorem 3. Let $\tilde{\phi} : M \times M \to G$ be a differentiable map satisfying (a) and (b) above. Then there exists a principal fiber bundle over M with connection Γ, unique up to isomorphism, such that $\tilde{\phi}$ is the restriction of the holonomy map of Γ at x to $\mathcal{S}_x$.

III. Gauge Fields and Gravity.

It should be noted that in a gauge theory there is a canonical metric, that is used to define the action which is invariant under gauge transformations. This metric can be used to determine the geodesics used in the definition of $\tilde{\phi}$ in section II, provided there exists a unique geodesic through any pair $y, z \in M$. If the last condition is not met then theorem 3 can still be proved locally.

For the gravitational field, G in section II may he chosen to be GL(4). However, to describe the gravitational field it is then necessary to introduce a solder form, in the GL(4) bundle [4] that is determined by ϕ up to isomorphism by theorems 1 and 2. It is known that for every linear connection, the solder form determines a unique affine connection [5]. We can therefore also describe the gravitational field by a metric and a holonomy map $\phi: \mathcal{L}_x \to A(4)$, where A(4) is the affine group in four dimensions. If metric compatibility of the connection is required then the image of ϕ is a subgroup of the Poincare group.

Hence the gravitational field and gauge field connections can be described in a unified manner by specifying the homomorphism $\phi: \mathcal{L}_x \to P \times K$ where P is the Poincare group and K is the gauge group. This suggests a unified classification of gravity and gauge fields in terms of the image of ϕ which is a subgroup of $P \times K$.

IV. Newtonian Gravity as a Gauge Field.

As already mentioned in section I, the phase shift in quantum interference provides a method of determining the holonomy map ϕ of the gravitational or gauge field that causes this phase shift.

More generally speaking, I shall call the holonomy map determined by the quantum mechanical motion of particles the quantum holonomy map, in order to distinguish it from the holonomy map determined by the classical motion of particles, which I shall call the classical holonomy map.

For instance, in general relativity, the classical holonomy map is a map into the Poincare group, whereas the quantum holonomy map is a map into the covering group of the Poincare group. It is known that classical Newtonian gravity can be given a space-time description with the gravitational field being described by a linear connection [6]. Thus Newtonian gravity is like other gauge fields in that the dynamical role is played by the connection, while the metric is fixed. The (classical) holonomy map ϕ corresponding to this connection is a map into the group of Galilei boosts such that ϕ maps curves that lie on constant time hyper-surfaces into the identity element.

Suppose now that the quantum holonomy map for Newtonian gravity is the same as the classical holonomy map. Then Schrodinger's equation in the absence of gravity must be modified in the presense of gravity by replacing the time derivative $\frac{\partial}{\partial t}$ by $\frac{\partial}{\partial t} + iA_0^i(\vec{x}, t)\, B^i$, where B^i generate local Galilei boosts, while the spatial derivatives should not be modified. It is known that $B^i = \dot{m}x^i$ where x^i are position operators. This implies that the Schrodinger's equation in the presence of gravity must be of the form $i\hbar\frac{\partial\psi(\vec{x}, t)}{\partial t} - mV(\vec{x}, t) = -\frac{\hbar^2}{2m}\nabla^2\psi\,(\vec{x}, t)$. We conclude by noting that the treatment of gauge fields by holonomy maps seems to be more general than the usual treatment of gauge fields in that it enables us to regard even the Newtonian potential in Schrodinger's equation as a gauge field.

Acknowledgements.

I wish to thank D. Friedan, T. Jacobson and I.M. Singer for useful and stimulating discussions. This work is based on seminars I delivered at the University of Maryland during September 1980 and at the University of Texas, Austin, during October 1980 and I wish to thank these institutions for financial support.

References.

1. J. Anandan, Il Nuovo Cimento, 53A, 221(1979); J. Anandan in Quantum Theory and Gravitation, edited by A.R. Marlow(Academic Press, 1980).

2. S. Kobayashi, Comptes Rendus, 238, 318, 443(1954).

3. J. Anandan, Int. J. Theor. Phys. 19, 537(1980), appendix.

4. A. Trautman in The Physicist's Conception of Nature, edited by J. Mehra (Reidel, Holland, 1973).

5. S. Kobayashi and K. Nomizu, Foundations of Differential Geometry (Interscience, 1969), Ch. III.

6. See, for instance, A. Trautman in Perspectives in Geometry and Relativity, edited by B. Hoffmann (Indiana Univ. Press, London 1966), p. 413.

How Spatial Topology Affects the Yang-Mills Vacuum

G. Kunstatter

Physics Dept.
University of Toronto
Toronto, Ontario
Canada, M5S 1A7

In the standard picture of the non-perturbative Q.C.D. vacuum [1], the gauge potentials are restricted by strong boundary conditions which compactify three space from R^3 to S^3. Consequently, the group of (time-independent) gauge transformations is disconnected. That is,

$$\pi_o(SU3^{S^3}) = [S^3, SU3] = \pi^3(SU3)$$
$$= Z \,, \qquad (1)$$

where we have used the notation Y^X to denote the space of pointed maps from X into Y, and [X,Y] to denote the set of homotopy classes of these maps. Since the space of zero energy solutions to the field equations on R^3 is in bijective correspondence with the group of gauge transformations, gauge fixing does not fix the vacuum uniquely. There exists a countable infinity of degenerate quantum "n-vacua" peaked around homotopically in-equivalent classical pure gauge configurations. These n-vacua are not invariant under homotopically non-trivial gauge transformations. Moreover, the existence of instanton solutions [2], which interpolate between the n-vacua in Euclidean space-time with finite action, implies that quantum mechanical tunneling can occur between the n-vacua. Thus the diagonalized θ-vacua

$$|\theta\rangle = \sum_{n\varepsilon Z} e^{in\theta}|n\rangle \qquad (2)$$

have non-degenerate θ-dependent energies.

The above scenario starts from a space-time which is assumed to be R^4. However, non-trivial spatial topologies can occur in quantum field theory for various reasons: complicated boundary conditions imposed on the fields, the presence of a classical gravitational field which affects the large-scale structure of space, and the possibility of quantum gravitational fluctuations at the Planck length which cause small scale fluctuations in topology. It is therefore natural to ask how the standard picture of the Yang-Mills vacuum is altered by the presence of non-trivial spatial topology. In the following, I will discuss briefly the results of an investigation [3] into the nature of the n-vacua in a canonically quantized Yang-Mills theory with gauge group G defined on an arbitrary compact three space, Σ. Details can be found in reference [3], while reference [4] contains a more heuristic description with physical examples.

In the temporal gauge (A_o=0), the zero energy solutions to the Yang-Mills equations describe flat connections on a principal G-bundle over Σ. When Σ is multiply-connected, these connections are not all pure gauge. In particular, there may exist flat connections on non-trivial G-bundles over Σ. The topological charges which label these inequivalent bundles will be conserved, because finite action transitions between inequivalent bundles cannot occur. For example, flat U1 bundles over Σ are labelled by the torsional elements of $H^2(\Sigma,Z)$. When $\Sigma=RP_3$, $H^2(RP_3,Z)=Z_2$ and there exist two distinct vacuum sectors.

Even if the G-bundle is trivial, when $\pi_1(\Sigma)\neq 0$ there may be flat connections which are not pure gauge globally. Such connections necessarily have a non-trivial discrete holonomy group. It was shown in reference [3] that they could have drastic consequences for the n-vacua in Yang-Mills theory. First, flat connections with non-zero holonomy group could give rise to new vacuum sectors. Although tunneling is expected between such holonomy sectors, no calculation of the transition amplitudes has yet been made. An example in which new sectors appear is provided by SU2 Yang-Mills theory over RP_3. Here, in addition to the usual countable infinity of n-vacua peaked around pure gauge potentials, there exist an additional integer's worth of vacua peaked around connections with holonomy group Z_2.

Alternatively, a path of connections with non-trivial holonomy group might interpolate with zero energy between different homotopy classes of pure gauge configurations. This could lead to a collapse of the corresponding n-vacua, since the tunneling amplitude would be effectively unity in the semi-classical approximation. Such a collapse of n-vacua occurs in the trivial example of a U1 gauge theory defined on S^1. In fact, for an Abelian gauge theory any two flat connections can be linked by a path of flat connections simply by taking the affine sum.

The results of the investigation outlined above suggest that spatial topology plays an important role in the vacuum structure of Yang-Mills theories and in quantum field theory in general. It is possible that further theoretical investigation along these lines will one day yield experimental predictions which would allow us to probe the large and small scale topological structure of the space-time in which we live.

References

[1] R. Jackiw and C. Rebbi, Phys. Rev. Letts. 37, 179, 1976.
C.J. Callan, R.F. Dashen and D.J. Gross, Phys. Letts. 63B, 334, 1976.

[2] A.A. Belavin, A.M. Polyakov, A.S. Schwartz and Yu. S. Tyupkin, Phys. Lett. 59B, 85, 1975.

[3] C.J. Isham and G. Kunstatter, Phys. Letts. 102B, 417, 1981.
C.J. Isham and G. Kunstatter, J. Math. Phys. (in print).

[4] G. Kunstatter, in The Quantum Structure of Space and Time eds M. Duff and C.J. Isham, proceedings of the Nuffield Workshop on Quantum Gravity, Imperial College, 1981 (in print).

NON-LINEAR DIFFERENTIAL SEQUENCES IN GAUGE THEORY

J.F. Pommaret
Ecole Nationale des Ponts et Chaussées, France

ABSTRACT: Classical gauge theory deals with connections on principal bundles. Our purpose is to prove that it is much more powerful to deal with Lie groupoids and the non-linear Spencer sequences, though the passage from the first to the second Spencer sequence is breaking the use of connections. Such a dilemma must be solved by future physics and the differential geometry of gauge theory must be revisited.

If $\mathcal{P}$ is a principal bundle over the base manifold X with structure group $\mathcal{G}$, we have the short exact sequence of vector bundles pulled back over $\mathcal{P}$:

$$0 \longrightarrow V(\mathcal{P}) \longrightarrow T(\mathcal{P}) \longrightarrow T(X) \longrightarrow 0$$

and we define a connection χ as a splitting of the induced short exact sequence of vector bundles over X:

$$0 \longrightarrow V(\mathcal{P})/\mathcal{G} \longrightarrow T(\mathcal{P})/\mathcal{G} \overset{\chi}{\longleftarrow\!\!\longrightarrow} T(X) \longrightarrow 0$$

Using the same notation for a vector bundle and its sheaf of germs of sections with $T = T(X)$, we get the following non-linear sequence:

$$\mathcal{P} \longrightarrow T^* \otimes T(\mathcal{P})/\mathcal{G} \longrightarrow \wedge^2 T^* \otimes T(\mathcal{P})/\mathcal{G}$$

$$f \rightsquigarrow f^{-1} df \qquad \chi \rightsquigarrow \kappa$$

where $\kappa(\xi,\eta) = [\chi(\xi), \chi(\eta)] - \chi([\xi,\eta]), \ \forall \xi,\eta \in T$ is the curvature and the first bracket is induced by the natural bracket on $T(\mathcal{P})$.

The work of Spencer[3], which is not known by many physicists, has been to generalize the above results to more general situations.

Let $\Pi_q(X,X)$ be the bundle of q-jets of invertible maps from X to X and $\mathcal{R}_q \subset \Pi_q(X,X)$ be a non-linear involutive system of finite Lie equations, with solution sheaf Γ a Lie pseudogroup of transformations of X. We may introduce the q-jet $id_q = j_q(id)$ of the identity map in the vertical machinery and define the corresponding linear system

$R_q = id_q^{-1}(v(\mathcal{R}_q)) \subset J_q(T)$ of infinitesimal Lie equations.

When $\mathcal{R}_q$ is transitive, that is to say when the canonical map $\mathcal{R}_q \to X \times X$ is surjective, the kernel R_q^0 of the projection of R_q onto T is a bundle of Lie algebras and a connection χ_q of order q is now a splitting of the short exact sequence of vector bundles over X:

$$0 \longrightarrow R_q^0 \longrightarrow R_q \underset{\chi_q}{\overset{}{\rightleftarrows}} T \longrightarrow 0$$

Let us define a bracket { } on $J_{q+1}(T)$ by the formula:

$$\{j_{q+1}(\xi), j_{q+1}(\eta)\} = j_q([\xi,\eta]), \ \forall \ \xi,\eta \in T \ .$$

We may define a bracket [ ] on $J_q(T)$ by the formula:

$$[\xi_q, \eta_q] = \{\xi_{q+1}, \eta_{q+1}\} + i(\xi)D\eta_{q+1} - i(\eta)D\xi_{q+1}$$

for any sections ξ_{q+1}, η_{q+1} of $J_{q+1}(T)$ projecting onto the sections ξ_q, η_q of $J_q(T)$. We have introduced the Spencer operator $D: J_{q+1}(T) \to T^* \otimes J_q(T)$ and the standard interior multiplication $i(\ )$ as in Refs. 1 and 2.

One can prove that $[R_q, R_q] \subset R_q$ and exhibit the <u>first non-linear Spencer sequence</u> as follows:

$$0 \longrightarrow \Gamma \xrightarrow{j_{q+1}} \mathcal{R}_{q+1} \xrightarrow{\bar{D}} T^* \otimes R_q \xrightarrow{\bar{D}'} \wedge^2 T^* \otimes J_{q-1}(T)$$

where $\mathcal{R}_{q+1} = J_1(\mathcal{R}_q) \cap \Pi_{q+1}(X,X)$ is the first prolongation of $\mathcal{R}_q$,

and $\bar{D} f_{q+1} = f_{q+1}^{-1} \circ j_1(f_q) - id_{q+1}$, $\bar{D}\chi_q(\xi,\eta) = D\chi_q(\xi,\eta) + \{\chi_q(\xi), \chi_q(\eta)\}$, $\forall \xi, \eta \in T$

As a byproduct, if χ_q is a connection, then $\bar{D}'\chi_q$ is induced by the curvature $\kappa_q(\xi,\eta) = [\chi_q(\xi), \chi_q(\eta)] - \chi_q([\xi,\eta])$.
All the details are given in Ref. 2.

However, this is not the best sequence as neither $\bar{D}$ nor $\bar{D}'$ are involutive and we have to construct the <u>second non-linear Spencer</u>

sequence:

$$0 \longrightarrow \Gamma \xrightarrow{j_q} R_q \xrightarrow{\bar{D}_1} C_1 \xrightarrow{\bar{D}_2} C_2$$

where $C_r = \Lambda^r T^* \otimes R_q / \delta(\Lambda^{r-1} T^* \otimes G_{q+1})$, G_{q+1} is the symbol of R_{q+1} and δ is the map involved in the Spencer cohomology of the symbols.

A striking result for physicists is that the sections of C_1 that can be used must not be connections.

Moreover, when $q = 1$ and R_1 is the system of Killing equations for a metric, the latter sequence is just describing the structure of Cosserat media and its compatibility conditions.

References

1. J.F. Pommaret: Systems of partial differential equations and Lie pseudogroups, Gordon and Breach (1978).
2. J.F. Pommaret: Differential Galois theory, Gordon and Breach (1982).
3. D.C. Spencer, A. Kumpera: Lie equations, Annals. of Math. Studies 73, Princeton University Press (1972).

YANG-MILLS GAUGE AND THE GRAVITATIONAL GAUGE

D.K. Sen

Department of Mathematics
University of Toronto
Toronto, Canada M5S 1A1

Abstract: The problem of solving the Einstein-Yang-Mills equations for a combined gravitational and Yang-Mills field is considered as a purely geometrical problem of determining a linear connection on the principal frame bundle L(M) from a connection on a SU(2) principal bundle over a space-time M.

Introduction: According to the General Theory of Relativity, every non-gravitational field, via its energy-momentum tensor and the Einstein field equations, should generate a gravitational field. Thus, in the case of combined gravitational and Yang-Mills fields, geometrically, every connection on a SU(2) principal bundle over a space-time M should determine a linear connection on the principal frame bundle L(M) of M, whose structure group is GL(4). Since there exists the well-known 2-1 covering homomorphism of SU(2) into SO(3) and thus GL(4), it suggests that we consider homomorphisms of principal bundles. Then there exists a mapping theorem of connections on bundles whenever there is a diffeomorphism of the base manifolds.

The mapping theorem of connections on bundles: [1] Recall that a *bundle homomorphism* of a principal bundle (P,M,π,G) into another principal bundle (P',M',π',G') is a triple of C^∞-maps (f_P,f_G,f_M), where $f_P : P\to P'$, $f_M : M\to M$ and $f_G : G\to G'$ is also a group homomorphism and such that (i) $f_M \circ \pi = \pi' \circ f_P$ and (ii) $f_P \circ R_g = R_{f_G(g)} \circ f_P$ for all $g \in G$. Here R_g and $R_{f_G(g)}$ are the group actions on P and P' respectively. The mapping theorem of connections on bundles then states the following.

Theorem: Let (f_P,f_G,f_M) be a bundle homomorphism from (P,M,π,G) into (P',M',π',G') such that $f_M : M\to M'$ is now a *diffeomorphism*. Let Γ be a connection in P. Then there is a unique connection Γ' in P' such that the horizontal subspaces of Γ are mapped into the horizontal subspaces of Γ'.

Application to gauge fields: First of all, to see the relationship between Γ and Γ', let w, w' be the corresponding connection forms and Ω, Ω' the corresponding curvature forms of Γ, Γ' respectively. If now $\bar{G}$ and $\bar{G}'$ are the Lie algebras of G and G' respectively, the homomorphism $f_G : G\to G'$ induces a Lie algebra homomorphism $\bar{f}_G : G\to \bar{G}'$. Then according to the theorem

$$f_P^*(w') = \bar{f}_G \cdot w \tag{1}$$

and

$$f_P^*(\Omega') = \bar{f}_G \cdot \Omega . \tag{2}$$

(1) is to be interpreted as follows. For every $X_p \in T_p(P)$, $w'_{p'}\left((f_P)_{*p}(X_p)\right) \in \bar{G}'$, where $p' \in f_P(p)$. On the other hand, $w_p(X_p) \in \bar{G}$.

Hence (1) means

$$w'_{p'}\big((f_P)_{*p}(X_p)\big) = \bar{f}_G\big(w_p(X_p)\big)$$

for all p and X_p. Similarly for (2).

We shall now apply the mapping theorem to the following situation. We take the same base manifold: $M = M' = \mathbb{R}^4$ (for simplicity), and $P = P(M) = M \times SU(2) = \mathbb{R}^4 \times SU(2)$, i.e. the $SU(2)$-bundle over $\mathbb{R}^4$, and $P' = L(M) = \mathbb{R}^4 \times GL(4)$, the frame bundle of $\mathbb{R}^4$. Let now $f_M : \mathbb{R}^4 \to \mathbb{R}^4$ be any C^∞-map. There is a natural 2-1 covering homomorphism of $SU(2)$ into $SO(3) \subset GL(4)$. This gives us a homomorphism $f_G : SU(2) \to GL(4)$. If we now define $f_P : P \to P'$ by $(x,g) \to \big(f_M(x),\ f_G(g)\big)$ where $x \in \mathbb{R}^4$ and $g \in SU(2)$, it is easy to see that (f_P, f_G, f_M) is a bundle homomorphism of $P(M)$ into $L(M)$. If furthermore, f_M is a *diffeomorphism* of $\mathbb{R}^4$ onto $\mathbb{R}^4$, then the mapping theorem on connections says that a connection Γ on $SU(2)$-bundle $P(M)$ determines uniquely a linear connection Γ' on the frame bundle $L(M)$ via equations (1) and (2), such that the horizontal subspaces of $P(M)$ are mapped into the horizontal subspaces of $L(M)$.

Since $SU(2)$, as a manifold, is homeomorphic to S^3, we can parametrize it by means of 3 real parameters $u = (U_1, u_2, u_3) \in S^3 \simeq SU(2)$. Let
$f_G : SU(2) \to GL(4)$ be given by $u \mapsto a^\alpha_\beta(u) \in GL(4)$, so that
$f_P : \mathbb{R}^4 \times SU(2) \to \mathbb{R}^4 \times GL(4)$ is given by $(x,u) \mapsto \big(f_\lambda(x),\ a^\alpha_\beta(u)\big)$ where
$f_M : \mathbb{R}^4 \to \mathbb{R}^4$ is $(x_\lambda) \mapsto \big(f_\lambda(x)\big)$. We shall write the map as $f_P : (X_a) \mapsto (Y_A)$ where $(X_a) = (x,u)$ and $(Y_A) = \big(f_\lambda(x),\ a^\alpha_\beta(u)\big)$. A Yang-Mills form w on $P(M)$ is then $w(X) = w_a(X)dX_a$ whereas a linear connection form w' on $L(M)$ is $w'(Y) = w'_A(Y)dY_A$. Equation (1) is then simply

$$\frac{\partial Y_A}{\partial X_a} w'_A(Y) = \bar{f}_G\big(w_a(X)\big). \tag{3}$$

Now

$$w'(Y) = w'_A(Y)dY_A = \Big\{(a^{-1})^\beta_\nu(u)\,\big(da^\nu_\mu(u) + \Gamma^\nu_{\alpha\lambda}(y)a^\lambda_\mu(u)dy_\alpha\big)\Big\}E^\mu_\beta$$

where $\Gamma^\nu_{\alpha\lambda}(y)$ are the connection components on the base manifold $\mathbb{R}^4$ and $\{E^\mu_\beta\}$ are the standard basis of $\overline{GL(4)}$, and $w(X) = w_a(X)dX_a = \{w^i_a(X)dX_a\}E_i = \{w^i_j(u)du_j + w^i_\alpha(x,u)dx_\alpha\}E_i$ where $w^i_\alpha(x,u)$ are related to the Yang-Mills potentials and $\{E_i\}$ are the basis of $\overline{SU(2)}$. Thus from (3), we get

$$\begin{aligned}&\Big\{\big(a^{-1}\big)^\beta_\nu(u)\,\big(da^\nu_\mu(u) + \Gamma^\nu_{\alpha\lambda}(y)a^\lambda_\mu(u)dy_\alpha\big)\Big\}E^\mu_\beta \\ &= \bar{f}_G\Big(\big\{w^i_j(u)du_j + w^i_\alpha(x,u)dx_\alpha\big\}E_i\Big)\end{aligned} \tag{4}$$

Recall that $\bar{f}_G = (f_G)_{*e} : T_e\big(SU(2)\big) \to T_{f_G(e)}\big(GL(4)\big)$, $SU(2) \ni e \leftrightarrow (u=0)$. So we put $u=0$ above and let $\bar{f}_G(E_i) = C^{\beta}_{i\mu} E^{\mu}_{\beta}$. Then (4) becomes

$$
\begin{aligned}
&(a^{-1})^{\beta}_{\nu}(0)\left(\left.\frac{\partial a^{\nu}_{\mu}(u)}{\partial u_i}\right|_{u=0} du_i + \Gamma^{\nu}_{\alpha\lambda}\big(y(x)\big)a^{\lambda}_{\mu}(0)dy_{\alpha}\right) \\
&= \left\{w^i_j(0)du_j + w^i_{\alpha}(x,0)dx_{\alpha}\right\}C_{i\mu}^{\ \beta} .
\end{aligned}
\tag{5}
$$

But $a^{\lambda}_{\mu}(0) = (a^{-1})^{\lambda}_{\mu}(0) = \delta^{\lambda}_{\mu}$ and $w^i_j(0) = \delta^i_j$

$$
\left.
\begin{aligned}
\text{so} \qquad & C^{\beta}_{i\mu} = \left.\frac{\partial a^{\beta}_{\mu}(u)}{\partial u_i}\right|_{u=0} \\
\text{and} \qquad & \Gamma^{\beta}_{\lambda\mu}\big(y(x)\big)\frac{\partial y_{\lambda}}{\partial x_{\alpha}} = C^{\beta}_{i\mu} w^i_{\alpha}(x,0)
\end{aligned}
\right\} .
\tag{6}
$$

Eq. (6) thus determines the connection components $\Gamma^{\beta}_{\lambda\mu}$ in terms of the Yang-Mills potentials w^i_{α} and the functions $y_{\lambda} = f_{\lambda}(x)$.

They also show that the contribution of the Yang-Mills potentials to the connection components is, in general, non-symmetric and would thus give rise to *torsion* in the connection.

This viewpoint seems to suggest that, in analogy with Yang-Mills field, we should regard the $\Gamma^{\beta}_{\lambda\mu}$'s, instead of the metric, as potentials of the gravitational field and this lends some support to Yang and Killmister's [2] suggestion that the source-free field equations of gravitation should be of the form: $R_{\mu\alpha;\beta} = R_{\mu\beta;\alpha}$.

References:

[1] S. Kobayashi and K. Nomizu, "Foundations of Differential Geometry", Vol. I, Interscience, New York (1963).

[2] C.N. Yang, Phys. Rev. Lett. 33, 445 (1974).
C.W. Killmister, "Perspectives in Geometry and Relativity", Ed. Hoffman, Bloomington (1966).

TWISTED FIELD THEORIES

S.D. Unwin
Department of Theoretical Physics
The University, Newcastle upon Tyne
NE1 7RU, U.K.

Since it was originally suggested that one may generalise, in spaces with certain non-trivial topologies, the usual definition of tensor field to include 'twisted' fields, the direct physical interpretation of these structures has not been clear. They may satisfy identical local dynamical constraints as the corresponding untwisted fields and yet when incorporated into familiar local field theories, lead to quite different physical effects. Here, two situations are considered in which the incorporation of twisted field configurations into the theory is not only an alternative to the inclusion of the usual untwisted fields, but for reasons technical, and perhaps even 'philosophical', a preferable alternative.

Modern gauge theories of fundamental forces predict vacuum contributions to the cosmological constant which are typically many orders of magnitude greater than its observed upper limit. This demands cancellations between terms constituting the cosmological constant which are so precise that it appears we must appeal to anthropic explanations of such fine tuning effects. The consideration of twisted field theories may, however, suggest a rather different approach to an understanding of the small cosmological constant. The only constant twisted scalar field is that which vanishes globally and, consequently, in the symmetry broken phase of a real, scalar field theory, one may expect the vacuum contributions to the cosmological constant, or, more precisely, the cosmological field, to be space-dependent [1,2]. The inclusion of a simple Goldstone Lagrangian, $\mathcal{L} = \frac{1}{2}\partial_\mu\phi\partial^\mu\phi + \frac{1}{2}m^2\phi^2 - \frac{1}{2}\xi R\phi^2 - \frac{1}{4}\lambda\phi^4$, in the usual gravitational action, for example, leads to a cosmological field, Λ, given by [2]

$$\Lambda = F(\sigma)\Lambda_o + 2\pi F(\sigma)G_o\left[2(4\xi - 1)\partial_\alpha\sigma\partial^\alpha\sigma + 8\xi\sigma\Box\sigma - 2m^2\sigma^2 + \lambda\sigma^4\right]$$

where

$$F(\sigma) = (1 - 8\pi\xi G_o\sigma^2)^{-1}.$$

Here, R is the scalar curvature of the spacetime, m, λ, ξ, G_o, Λ_o are constants (the last two corresponding to the newtonian and cosmological constants in the symmetric phase of the field theory) and σ is the vacuum expectation value of ϕ for a particular metric, a back reaction calculation following. We may envisage a situation in which the cosmological field is typically of a microphysical value yet, of necessity, we inhabit an atypical region of the universe where the value of Λ is close to zero ($\lesssim 10^{-57}$ cm^{-2}).

The second situation considered [3] is that for which scalar electrodynamics may be modelled in terms of a dimensionally reduced five-dimensional, free, real scalar field theory of Lagrangian $\partial_i\phi\partial^i\phi - \alpha\phi^2$, where i = 0 - 4, upon a spacetime of underlying manifold $\mathbb{R}^4 \otimes S^1$. The charges and masses associated with the scalar fields of the dimensionally reduced theory are identified in a suitable manner with functions of α and the circumference, L, of the fifth dimensional circle, yielding, for realistic charge and mass values, a negative (tachyonic) α and an L of order 10^{-31} cm. Noting that $H^1(\mathbb{R}^4 \otimes S^1, \mathbb{Z}_2) \approx \mathbb{Z}_2$, where the cohomology group $H^1(M, \mathbb{Z}_2)$ counts the inequivalent real line bundles over the manifold M, we deduce that there may exist twisted, real scalar fields on the spacetime considered. It is discovered, as a consequence of the existence of complex eigenfrequencies, that the untwisted five-dimensional field theory is unstable, even at the tree graph level, whereas no such instabilities arise in the corresponding twisted field theory. The antiperiodicity conditions associated with the twisted scalar field put a lower bound on the momentum states available to it, thus ensuring that all the eigenfrequencies are real, and hence that the field theory is stable.

REFERENCES

[1] P.C.W. Davies and S.D. Unwin, 1981, Proc. R. Soc. A 377, 147.

[2] S.D. Unwin, 1982, J. Phys. A 15. 'Twisted Symmetry Breaking on the Projective Hypersphere: A Model of the Small Cosmological Constant'.

[3] S.D. Unwin, 1981, Phys. Lett. 103B, 18.

A DIFFERENTIAL CLASSIFICATION AND A QUASI-METRIC ASSOCIATED WITH SU(2) YANG-MILLS FIELD STRENGTHS

H.K. Urbantke
Institute for Theoretical Physics
University of Vienna

and

SISSA, Trieste

Abstract: By studying the plane elements on which parallel displacement of isovectors is integrable, we provide a differential classification of SU(2) Yang-Mills field strengths. In the generic case, a quasi-metric is naturally associated with the field. By studying the conditions under which the plane elements mentioned are tangent to submanifolds the classification can be refined.

Basic for the final solution of the instanton problem was the remark of R. Ward [1] that a gauge potential with selfdual field strength has a parallel displacement of isovectors that is integrable on the antiselfdual totally null 2-planes of (complex) Minkowski space. Virtually the same observation lies at the basis of Yang's approach [2] who uses these planes as coordinate planes to simplify the field equations.

In the case of SU(2) Yang-Mills field strengths that are not necessarily selfdual (in the sense of the Minkowski metric $\eta_{\mu\nu}$) we might nevertheless ask for those plane elements at each space time point x where parallel transport around infinitesimal loops yields the identity. The set of these plane elements, at each x, is of qualitatively different appearance for different algebraic types of the field strength $F^a_{\mu\nu}(x)$ (a, ... are SU(2) indices) and yields a classification of the latter which is quite coarse because it involves neither the Lorentz group nor the detailed SU(2) group structure; nevertheless it seems quite useful. Next one can investigate, in each case, the integrability conditions under which one can select plane elements such as to fit together to form surfaces on which parallel transport is integrable.

To find the plane elements, one can span them by two tangent vectors u^μ, v^μ, but what matters is only the antisymmetric product $p^{\mu\nu} = u^{[\mu}v^{\nu]}$ up to a non-zero scalar factor. The condition to be satisfied is $F^a_{\mu\nu}\ p^{\mu\nu} = 0$. The set of solutions depends on the rank r of the matrix $M^{ab} = F^a_{\mu\nu} F^b_{\alpha\beta}\ \varepsilon^{\mu\nu\alpha\beta}$ and can pictorially represented as a set of lines in 3-space (remember that 2-dimensional subspaces of a 4-dimensional vector space correspond to projective lines in projective 3-spaces). Using elementary methods of line geometry - a version readable by physicists being provided by Appendix III of Hlavaty's book [3] - one obtains the following classification.

<u>r = 3</u> ... lines form one system of generators on a quadric.

<u>r = 2</u> ... lines of a plane a through a point u and lines of a plane b through a point v, where $u \in a \cap b$, $v \in a \cap b$, $u \neq v$, $a \neq b$

<u>r = 1</u> ... the same, but $u = v \in a = b$.

<u>r = 0</u> ... lines of a plane or lines through a point.

The pictorial description easily translates to the physicist's space-time language. For bevity we have assumed the $F^a_{\mu\nu}$ to be linearly independent and do not discuss here the case of linearly dependent $F^a_{\mu\nu}$ nor the reality of the objects obtained. For r = 3, however, we find it noteworthy to point out that the quadric alluded to corresponds to a quasi-metric (determined up to a non zero factor)

$$g_{\mu\nu} = \varepsilon_{abc} F^a_{\mu\nu} F^b_{\rho\sigma} F^c_{\beta\nu} \varepsilon^{\alpha\rho\sigma\beta}$$

with inverse (up to a nonzero factor)

$$g^{\mu\nu} = \varepsilon_{abc} F^a_{\lambda\rho} F^b_{\alpha\beta} F^c_{\sigma\tau} \varepsilon^{\mu\alpha\lambda\rho} \varepsilon^{\nu\beta\sigma\tau}$$

with respect to which the $F^a_{\mu\nu}$ are selfdual while the plane element tensors $p^{\mu\nu}$ are antiselfdual (or vice versa); for real $F^a_{\mu\nu}$, $g_{\mu\nu}$ has signature ++++ or ++--.

Concerning the question whether one can select surface forming plane elements, it must be enough here to mention that in the r = 3 case a convenient parametrization of the plane elements is in terms of 2-component spinors of $g_{\mu\nu}$ and that a straight forward application of the Frobenius condition then leads to a condition on the Weyl tensor of $g_{\mu\nu}$ that will locally guarantee the surface-forming property. - We have not touched global questions.

The complete classification and its derivation, as well as the integrability results, will be published elsewhere. The first steps of this program were indicated in [4].

References

[1] R. Ward, Phys. Lett. 61A, 81 (1977)

[2] C.N. Yang, Phys. Rev. Lett. 38, 1377 (1977)

[3] V. Hlavatý, Geometry of Einstein's Unified Field Theory. P. Noordhoff, Groningen 1958.

[4] H. Urbantke, Acta Phys. Austr., Suppl. XIX, 875 (1978).

TOPOLOGICAL CHARGES IN SUPERSYMMETRIC YANG-MILLS THEORIES

P.A. Zizzi

International School for Advanced Studies, Trieste, Italy

It is well known that in supersymmetric theories the graded Lie algebra can be modified by the presence of central charges [1].

In the case of supersymmetric theories with solitons, one of the possible central charges is of topological origin. For supersymmetric Yang-Mills theories in particular, this fact has been shown in 3 + 1 dimensions by Olive and Witten [2] and in 4 + 1 dimensions by the author [3].

Olive and Witten considered the N = 2 supesymmetric O(3) Yang-Mills theory in 3 + 1 dimensions. They found that, in the phase of the theory where the gauge group O(3) isbroken down to a subgroup $U(1)$, the supersymmetry algebra is modified by the presence of two central charges. The first one is the electric charge E of the abelian subgroup $U(1)$, the second one is the topological charge G of the solitons (the 't Hooft Polyakov monopoles, pseudoparticles in 2 + 1 dimensions):

phase $\langle \phi \rangle \neq 0$

$$\{Q_{\alpha i}, \overline{Q}_{\beta j}\} = \delta_{ij} (\gamma^{\mu})_{\alpha\beta} P_{\mu} + \varepsilon_{ij} \left(\delta_{\alpha\beta} E + (\gamma_5)_{\alpha\beta} G\right) \tag{1}$$

with: $\alpha, \beta = 1,2,3,4 \quad ; \quad i,j = 1,2 \quad ; \quad \mu = 0,1,2,3$

where:

$$E = \frac{1}{\langle \phi \rangle} \int d^3x \ \partial_i \left(\phi^a F_{0i}^{a}\right) \tag{2}$$

$$G = \frac{1}{\langle \phi \rangle} \int d^3x \ \partial_i \varepsilon_{ijk} \left(\phi^a F_{jk}^{a}\right) \tag{3}$$

with: $i, j, k = 1,2,3 \quad ; \quad a = 1,2,3$.

The presence of the central charges E and G in the supersymmetry algebra (1) assures that the mass formula:

$$M = \langle \phi \rangle \sqrt{E^2 + G^2} \tag{4}$$

is valid not only at the classical level, but survives quantization. (M is the mass acquired by elementary particles and solitons through the Higgs mechanism). In the 3 + 1 dimensional case, both the electric charge and the topological charge are the usual central charges defined by Fayet [4] as the generators of the unbroken abelian subgroup.

In a recent paper [3], the results of Olive and Witten were extended to the N = 2 supersymmetric SU(2) Yang-Mills theory in 4 + 1 dimensions . It was found that the topological charge of the solitons (the BPST instantons in the 4-dimensional Euclidean space) modifies the supersymmetry algebra in the phase of the theory where the gauge group SU(2) is unbroken:

phase $\langle \phi \rangle = 0$

$$\{ \bar{Q}^i_\alpha , Q^j_\beta \} = \Omega^{ij} \left[(\gamma^\mu)_{\beta\alpha} P_\mu + \delta_{\alpha\beta} Z_T \right] \tag{5}$$

with: $\alpha, \beta = 1,2,3,4 \quad ; \quad i, j = 1,2 \quad ; \quad \mu = 0,1,2,3,4$

where Ω^{ij} is the symplectic metric used to raise and lower indices, and:

$$Z_T = -\frac{1}{4} \int d^4x \, F^a_{\ell m} \tilde{F}^{\ell m a} = \frac{8\pi^2}{g^2} q \tag{6}$$

with: $\ell, m = 1,2,3,4 \quad ; \quad a = 1,2,3$

Z_T is the topological charge of the solitons and q is the topological number of instantons. Because of the self-duality of instantons, the mass formula holds

$$M_{soliton} = \frac{Z_T}{q} \tag{7}$$

In this phase the elementary particles remain massless; owing to the absence of the Higgs mechanism.

In this case the definition of the central charge given by Fayet does not apply to the topological charge.

The electric charge modifies the supersymmetry algebra in the phase of the theory where the gauge group SU(2) is broken down to a subgroup $U(1)$:

Phase $\langle\phi\rangle \neq 0$

$$\{ \bar{Q}^{i}_{\alpha}, Q^{j}_{\beta} \} = \Omega^{ij} \left[(\gamma^{\mu})_{\beta\alpha} P_{\mu} + \delta_{\alpha\beta} Z_E \right] \tag{8}$$

where:

$$Z_E = \frac{1}{\langle\phi\rangle} \int d^4x \, \partial_\ell \left(\phi^a F^a_{0\ell} \right)$$

with : $\ell = 1,2,3,4$; $a = 1,2,3$

Z_E is the electric charge of the unbroken abelian subgroup $U(1)$.

In this phase all the elementary particles acquire a mass:

$$M = \langle\phi\rangle Z_E \tag{9}$$

given by the Higgs mechanism.

References

[1] J.R. Haag, J.T. Łopuszanski and M. Sohnius, Nucl. Phys. B88 (1975) 257.

[2] E. Witten and D. Olive, Phys. Lett. 78B (1978) 97.

[3] P. Zizzi, Nucl. Phys. B189 (1981) 317.

[4] P. Fayet, Nucl. Phys. B149 (1979) 137.

V. GRAVITY

ENERGY DEFINITION AND STABILITY FOR $\Lambda \neq 0$ GRAVITY

Stanley Deser

CERN, Geneva, Switzerland

and

Brandeis University, Waltham, MA 02254, USA

Abstract. The definition of energy and its use in studying stability in general relativity are extended to the case when there is a nonvanishing cosmological constant Λ. Existence of energy is first demonstrated for any model (with arbitrary Λ). It is defined with respect to sets of solutions tending asymptotically to any background space possessing time-like Killing symmetry, and is both conserved and of flux integral form. When $\Lambda<0$, the energy is shown to be positive, and hence stability established, for all systems tending to anti-De Sitter space. Supergravity methods are used here, upon defining spinorial charges as flux integrals which obey the global graded algrebra. For $\Lambda>0$, small excitations about De Sitter space are stable inside the event horizon. Outside excitations can contribute negatively due to the Killing vector's flip at the horizon. This is a universal phenomenon associated with the possibility of Hawking radiation. Apart from this effect, the $\Lambda>0$ theory appears to be stable, also at the semi-classical level.

1. INTRODUCTION

The observed smallness of the cosmological constant Λ, or equivalently of the vacuum energy density of the Universe is one of the major problems in current physics. It is highly unnatural from the point of view of particle physics since it demands extreme fine tuning of parameters. A possible way to exclude the cosmological constant would be to show that it leads to some fundamental classical or semiclassical instabilities in the Einstein theory. This was one motivation of the present work; related to it was the challenge of extending to the $\Lambda\neq0$ case the energy concepts so important when $\Lambda=0$ and more generally of understanding when conserved quantities may be defined in general relativity. The results have been obtained in collaboration with L. F. Abbott and details may be found in a forthcoming joint paper (Abbott and Deser 1981).

Stability of a bounded matter system in flat space is usually established by showing it to have positive energy with respect to a lowest, vacuum, state. For gravity, (with $\Lambda=0$) energy of any asymptotically flat solution is also perfectly definable with respect to flat space as vacuum. It turns out that this energy is always positive and that the theory (also in the presence of positive energy matter) is stable; quite general and rigorous results have been obtained in recent years (Brill & Deser 1968; Deser & Teitelboim 1977; Grisaru 1978;

Schoen & Yau 1979b,c; Witten 1981). On the other hand, when $\Lambda \neq 0$, flat space is no longer an acceptable background (since it does not solve the Einstein equations), but must be replaced as vacuum by the "flattest", maximally symmetric solutions of the cosmological equations, namely De Sitter O(4,1) or anti-De Sitter O(3,2) space according to whether $\Lambda>0$ or $\Lambda<0$, respectively. At this point, a number of problems arise for the stability programme. First, can any reasonable physical substitute for energy be defined? Having lost asymptotic Poincare invariance, one is left with the asymptotic De Sitter or anti-De Sitter algebra at infinity, for which P_μ^2 is no longer a Casimir operator, being replaced by the five-dimensional "rotations" $J_{ab}=-J_{ba}$ $(a,b = 0,\ldots,4)$. One must therefore show that J_{04}, which become P_0 upon contraction $(\Lambda \rightarrow 0)$, is acceptable and that this quantity is really definable as a flux integral at infinity so as to satisfy the asymptotic global algebra. This is accomplished in Section 2, through a general analysis of how to obtain and interpret conserved quantities with respect to a background which possesses symmetries. We will see that, associated to every such symmetry, there is a generator which is conserved, background-covariant, and of flux integral form. The preferred one, which we call the Killing energy, is that which is connected with a timelike Killing vector or symmetry. Fortunately, for both signs of Λ, the maximally symmetric backgrounds have timelike symmetries, associated with J_{04}. The other nine generators are not timelike (nor are P_i, $J_{\mu\nu}$ when $\Lambda=0$). A brief review of the properties of De Sitter spaces in Section 3 will show that for $\Lambda>0$, the necessary presence of an event horizon, where the timelike Killing vector becomes null and then spacelike requires a more careful analysis of stability. Within the horizon, however, and in all space for $\Lambda<0$ (where there is no horizon) it will be seen in Section 4 that small oscillations about vacuum have positive energy. The possiblity of negative energy for excitations outside the horizon (for $\Lambda>0$) is a reflection, in Hamiltonian form, of the generic features whih lead to Hawking radiation (Gibbons & Hawking 1977). In Section 5, we shall discuss semi-classical stability, against quantum mechanical tunnelling, for $\Lambda>0$, which would be violated in the presence of Euclidean "bounce" solutions (Coleman 1977; Coleman & Callan 1977; Perry 1981; Gross, Perry & Yaffe 1981; Witten 1981b); no evidence for these is found. Section 6 is devoted to a demonstration of stability for all asymptotically anti-De Sitter solutions when $\Lambda<0$ by showing that the full energy is positive in that case. Here one uses methods of supergravity, parallel to those which were used to establish (Deser & Teitelboim 1977; Grisaru 1978) positivity of the energy for $\Lambda=0$. To accomplish this it is first necessary to define spinorial charges which are conserved and also have flux integral form and show the corresponding existence of appropriate Killing spinors, whose presence is implicit but rather trivial when $\Lambda=0$.

We conclude that models with cosmological constant are quite similar, in their energy definition and consequent stability properties, to the usual $\Lambda=0$ case. [This is not to say that they do not have other, unrelated, idiosyncrasies (Hawking & Ellis 1973).] Thus, although we have failed to exclude $\Lambda\neq 0$ on stability grounds, a unification has been achieved in our understanding both of these models and of the general Hamiltonian mechanism underlying the modifications in stability when event horizons are present.

2. CONSERVED QUANTITIES

Consider the physical system defined by the Einstein equations

$$G_{\mu\nu} + \Lambda g_{\mu\nu} = 0 \tag{2.1}$$

together with a background metric $\bar{g}_{\mu\nu}$ which satisfies (2.1); we decompose the full metric $g_{\mu\nu}$ according to

$$g_{\mu\nu} = \bar{g}_{\mu\nu} + h_{\mu\nu} \tag{2.2}$$

where $h_{\mu\nu}$ is not necessarily small, but does obey the boundary condition that it vanishes asymptotically at some appropriate speed. We will construct conserved quantities from $(\bar{g}_{\mu\nu},h_{\mu\nu})$ corresponding to the symmetries of the background. Although we are primarily concerned with background De Sitter or anti-De Sitter spaces, the method is completely general, and leads to flux integral expressions for these generators, which are constructed from the gravitational stress-tensor and the appropriate Killing vectors. Our conventions are that $R_{\mu\nu} \equiv R^{\alpha}{}_{\mu\alpha\nu} \sim + \partial_{\alpha}\Gamma^{\alpha}{}_{\mu\nu}$, signature $(+++-)$ and all operations such as covariant differentiation $(\bar{D}_{\mu})$ or index moving are with respect to $\bar{g}_{\mu\nu}$. We define the symmetric stress tensor $T^{\mu\nu}$ to be all terms of second and higher order in $h_{\mu\nu}$ when the decomposition (2.2) is inserted in (2.1):

$$G_L^{\mu\nu}(\bar{g},h) + \Lambda h_{\mu\nu} = T^{\mu\nu} = T^{\nu\mu} \quad . \tag{2.3}$$

The subscript L refers to terms linear in $h_{\mu\nu}$. Using the fact that $G_{\mu\nu}(\bar{g}) + \Lambda\bar{g}_{\mu\nu} = 0$ and that the left side of (2.3) therefore obeys the (exact) linearized Bianchi identity $\bar{D}_{\mu}(G^{\mu\nu} + \Lambda h^{\mu\nu}) = 0$, the field equations imply covariant conservation of $T^{\mu\nu}$:

$$\bar{D}_{\mu}T^{\mu\nu} = 0 \ . \tag{2.4}$$

To turn covariant into ordinary conservation, we have to define a conserved (background) contravariant vector density J^{μ} from the tensor $T^{\mu\nu}$ since then, and only then, is $\bar{D}_{\mu}J^{\mu} \equiv \partial_{\mu}J^{\mu}$. When $\bar{g}_{\mu\nu}$ has a symmetry, there exists a Killing vector $\bar{\xi}_{\mu}$, obeying

$$\bar{D}_{\mu}\bar{\xi}_{\nu} + \bar{D}_{\nu}\bar{\xi}_{\mu} = 0 \ . \tag{2.5}$$

Consequently,

$$\bar{D}_\mu(T^{\mu\nu}\bar{\xi}_\nu) = (\bar{D}_\mu T^{\mu\nu})\bar{\xi}_\nu + 1/2\ T^{\mu\nu}(\bar{D}_\mu\bar{\xi}_\nu + \bar{D}_\nu\bar{\xi}_\mu) = 0 \quad (2.6)$$

and $\sqrt{-\bar{g}}T^{\mu\nu}\bar{\xi}_\nu$ is the desired contravariant tensor density, for which true conservation holds:

$$\bar{D}_\mu(\sqrt{-\bar{g}}T^{\mu\nu}\bar{\xi}_\nu) \equiv \partial_\mu(\sqrt{-\bar{g}}T^{\mu\nu}\bar{\xi}_\nu) = 0\ . \quad (2.7)$$

So to every Killing vector, there is associated a conserved generator

$$E(\bar{\xi}) = 1/8\pi G \int d^3x\sqrt{-\bar{g}}T^{o\nu}\bar{\xi}_\nu \quad (2.8)$$

In particular, if $\bar{\xi}_\nu$ is timelike, this is the Killing energy.

Despite the fact that $\Lambda\neq 0$, we now show that $E(\bar{\xi})$ can be written in flux integral form, just as for $\Lambda=0$. From (2.3) it follows that

$$T^{\mu\nu} \equiv \bar{D}_\alpha\bar{D}_\beta K^{\mu\alpha\nu\beta} + 1/2(\bar{R}^\mu{}_{\alpha\beta}{}^\nu H^{\alpha\beta} - \Lambda H^{\mu\nu})\ . \quad (2.9)$$

Here the superpotential K is defined by

$$2K^{\mu\alpha\nu\beta} \equiv \bar{g}^{\mu\beta}H^{\nu\alpha} + \bar{g}^{\nu\alpha}H^{\mu\beta} - \bar{g}^{\mu\nu}H^{\alpha\beta} - \bar{g}^{\alpha\beta}H^{\mu\nu},$$

with

$$H^{\mu\nu} \equiv h^{\mu\nu} - 1/2\ \bar{g}^{\mu\nu}\ h^\alpha{}_\alpha\ . \quad (2.10)$$

It has the algebraic symmetries of the Riemann tensor:

$$K^{\mu\alpha\nu\beta} = -K^{\alpha\mu\nu\beta} = K^{\alpha\mu\beta\nu} = K^{\nu\beta\mu\alpha}\ . \quad (2.11)$$

Symmetry and conservation of $T^{\mu\nu}$ in (2.9) can easily be checked using the background field equations and its derivative consequences,

$$\bar{D}_\beta R_{\mu\nu\alpha}{}^\beta \equiv \bar{D}_\nu\bar{R}_{\alpha\mu} - \bar{D}_\mu\bar{R}_{\alpha\nu} = 0\ , \quad \bar{D}_\beta\bar{R}^{\mu\nu} = 0$$

but without any assumptions on the full background Riemann tensor.

Next, we form $T^{\mu\nu}\bar{\xi}_\nu$, and recast it into the expression

$$\sqrt{-\bar{g}}\ T^{\mu\nu}\bar{\xi}_\nu = \sqrt{-\bar{g}}\ \bar{D}_\alpha\left[(\bar{D}_\beta K^{\mu\alpha\nu\beta})\bar{\xi}_\nu - K^{\mu\beta\nu\alpha}\bar{D}_\beta\bar{\xi}_\nu\right]$$
$$\equiv \bar{D}_\alpha F^{\mu\alpha}\ . \quad (2.12)$$

It may be verified that all additional terms "miraculously" vanish upon use of the Killing identity $\bar{D}_\beta\bar{D}_\alpha\bar{\xi}_\nu + \bar{R}^\lambda{}_{\beta\alpha}{}^\nu\bar{\xi}_\nu \equiv 0$. Furthermore the quantity $F^{\mu\alpha}$ is (almost obviously) an antisymmetric tensor density, and therefore its divergence is an ordinary one, $\bar{D}_\alpha F^{\mu\alpha} \equiv \partial_\alpha F^{\mu\alpha}$. Hence the desired result:

$$8\pi GE(\bar{\xi}) = \int d^3x\sqrt{-\bar{g}}\; T^{o\nu}\xi_\nu = \oint dS_i F^{oi} \ . \qquad (2.13)$$

Note also that the Killing generator is mainfestly background-covariant, i.e. independent of the coordinate choice used for the background metric. As a check, when $\Lambda=0$ and the background is chosen to be flat, introduction of Cartesian coordinates simplifies (2.13) to be the usual expression for the Poincare generators. In particular, when $\bar{\xi}_\mu = (1,0)$ we obtain the standard energy formula

$$16\pi G\ E = \oint\ dS_i(h_{ji,j} - h_{jj,i}) \qquad (2.14)$$

which correctly reproduces the mass of any asymptotically Schwarzschild metric. That the ten Poincare generators obey the (global) Poincare algebra is an immediate consequence of the commutation properties of the Killing vectors.

When $\Lambda\neq0$, with $\bar{g}_{\mu\nu}$ a De Sitter or anti-De Sitter metric, we would obtain the 10 Killing generators corresponding to the background De Sitter or anti-De Sitter symmetries, and they automatically satisfy the appropriate global algebra. In particular, we get the timelike $E(\bar{\xi})$ expression by using the appropriate Killing vectors of Section 3. [A check here is to verify that the equivalent of the Schwarzschild solution for $\Lambda\neq0$ has energy m. For $\Lambda>0$, there are corrections because one must stay within the unavoidable event horizon (rather than go to infinity) in calculating the flux integral; apart from these, the correct result E=m emerges.] We also mention that this whole procedure could also have been carried out in first order form (used in Section 4) to yield $E(\bar{\xi})$ there as well.

We complete this section with a treatment of the graded algebra which can be introduced when $\Lambda<0$ (but not $\Lambda>0$!), in terms of spinorial charges Q. The resulting local supersymmetry is that of supergravity with a cosmological term (Townsend 1977) and a spin 3/2 "mass" term (Deser & Zumino 1977). This will be used in Section 6 to show that the supergravity energy operator is positive and from this establish stability for classical gravity with $\Lambda<0$. First, however, we must show that the spinor charges can be written as surface integrals as in (2.13) so as to satisfy the graded global algebra at infinity. The spinorial charge density is

$$Q^\mu = \varepsilon^{\mu\alpha\beta\nu}\bar{\gamma}_5\gamma_\alpha\tilde{D}_\beta\psi_\nu \quad . \qquad (2.15)$$

Its origin may be understood, just like that of $T^{\mu\nu}$, in terms of a decomposition of the full Rarita-Schwinger equation into a "linear" part and a remainder. Here ψ_ν is the spin 3/2 field, $\bar{\gamma}_\alpha$ are the background covariant γ matrices with respect to the background vierbein and the modified covariant derivative on a spinor, $\tilde{D}_\beta$, defined by

$$\tilde{D}_\beta \equiv \bar{D}_\beta + \frac{1}{2} m \bar{\gamma}_\beta \quad , \quad m^2 \equiv \frac{1}{3} |\Lambda| \tag{2.16}$$

has the basic property that $[\tilde{D}_\beta,\tilde{D}_\alpha]=0$ for a background anti-De Sitter space. The current Q^μ satisfies $\tilde{D}_\mu Q^\mu = 0$, and to convert this to an ordinary conservation law, we introduce Killing spinors obeying $\tilde{D}_\mu\alpha=0$ (consistent with $[\tilde{D}_\nu,\tilde{D}_\mu]\alpha=0$). The quantity $\bar{\alpha}Q^\mu$ is easily seen to take the form

$$\bar{\alpha}Q^\mu = \bar{D}_\beta(\bar{\alpha}\varepsilon^{\mu\alpha\beta\nu}\gamma_5\bar{\gamma}_\alpha\psi_\nu) \equiv \partial_\beta(\bar{\alpha}\varepsilon^{\mu\alpha\beta\nu}\gamma_5\bar{\gamma}_\alpha\psi_\nu) \ . \tag{2.17}$$

The last equality follows because the quantity in parentheses is an antisymmetric tensor density. But since $\bar{\alpha}Q^\mu$ is a contravariant vector, we see immediately that its ordinary divergence vanishes identically, $\partial_\mu(\bar{\alpha}Q^\mu)\equiv 0$, so that the spinor charge is both conserved and has the flux form

$$Q(\alpha) \equiv \int d^3x\bar{\alpha}Q^\circ \equiv \oint dS_j\bar{\alpha}\varepsilon^{\circ ijk}\gamma_5\bar{\gamma}_i\psi_k) \tag{2.18}$$

This is the required analogue of (2.13) for the bosonic generators. [The analogy actually goes further, in that only one "Coulomb" component of ψ_k enters in (2.18), corresponding to the "Coulomb" part of the metric in (the appropriate generalization of) Eq. (2.14)].

Each of the four independent Killing spinors $\alpha_{\beta'}{}^{(\beta)}$, where β' is the spinor index and (β) is the label of each spinor, defines a fermionic charge $Q_{(\beta)}$. These then satisfy the anticommutation relation

$$\{Q_{(\beta)},\bar{Q}_{(\beta')}\} = \frac{1}{2}\left(\gamma^{(\mu)}\right)_{(\beta\beta')} J_{(\mu 4)} + \left(\sigma^{(\mu\nu)}\right)_{(\beta\beta')} J_{(\mu\nu)} \tag{2.19}$$

if we take the α's to commute (for convenience). Having defined the required conserved quantities, we turn next to the choice of Killing vectors.

3. DE SITTER SPACES

We give a brief review of the symmetries of "vacuum" spaces when $\Lambda\neq 0$. De Sitter space corresponds to a four-surface $z_\mu^2 + z_4^2 = 3/\Lambda$, $\Lambda>0$, in flat five-space. Among the rotations of the embedding space are the boosts mixing z_o with (z_i,z_4). For example $\xi_a = (z_4,0,0,0,z_o)$ is a timelike Killing vector when $|z_4| > |z_o|$, but signals the existence of an event horizon at $|z_4| = |z_o|$, where stability must be discussed separately, since $E(\xi)$ no longer acts like an energy beyond it. Of course, an observer will only interact with events inside the horizon, which means that $E(\bar{\xi})$ tests stability to excitations visible to the observer. It is illumination to apply these ideas to a simple model, namely a scalar field in De Sitter space. Representing the metric in the form

$$ds^2 = -dt^2 + f^2(t)(dx^2 + dy^2 + dz^2) \ , \quad f(t) \equiv \exp\sqrt{\tfrac{\Lambda}{3}}\, t \qquad (3.1)$$

with the "timelike" vector

$$\xi^\mu = (-1, \sqrt{\tfrac{\Lambda}{3}}\,\vec{x}) \quad , \quad \xi^2 = -1 + \frac{\Lambda}{3}\,|\vec{x}f|^2 \qquad (3.2)$$

we see that the horizon appears (at any given time) for distances such that $|\sqrt{\Lambda/3}\ \vec{x}f| = 1$. The action and energy momentum densities for this theory are

$$I = \int d^4x f^3[\tfrac{1}{2}\dot{\phi}^2 - \tfrac{1}{2}f^{-2}(\nabla\phi)^2 - V(\phi)]$$

$$-T^o{}_o = \frac{1}{2}\, f^{-3}\pi^2 + \frac{1}{2}\, f(\nabla\phi)^2 + f^3V(\phi) \qquad (3.3)$$

$$-T^o{}_i = \pi\partial_i\phi$$

where $\pi \equiv f^2\dot{\phi}$. The energy density is positive [if $V(\phi)$ is] but time-dependent. The conserved energy (which is of course not a flux integral here) is still $E(\bar{\xi}) = \int d^3x\ T^{0\mu}\bar{\xi}_\mu$. The integrand has the form

$$T^{o\mu}\bar{\xi}_\mu = \{\frac{1}{2}\,[f^{-3}\pi^2 + f(\nabla\phi)^2] - \sqrt{\tfrac{\Lambda}{3}}\vec{x}\cdot\pi\nabla\phi\} + f^3V(\phi) \ . \qquad (3.4)$$

It will be positive provided the bracketed quantity is. The triangle inequality

$$\frac{1}{2}(\vec{A}^2+\vec{B}^2) > (\sqrt{\tfrac{\Lambda}{3}}f|\vec{x}|)\vec{A}\cdot\vec{B} \quad , \ \vec{A} \equiv f^{-3/2}\pi\hat{x}, \ \vec{B} \equiv f^{1/2}\vec{\nabla}\phi \ , \ \sqrt{\tfrac{\Lambda}{3}}f|\vec{x}| < 1 \qquad (3.5)$$

makes it easy to see that the positivity condition corresponds to $\bar{\xi}^2<0$ in (3.2), i.e., to excitations within the horizon. This correlation between the event horizon and positivity is an expression in Hamiltonian form of Hawking radiation (Gibbons & Hawking 1977), and will be seen later to be universal.

Anti-De Sitter space is the covering space for the surface $z_\mu^2 + z_4^2 = 3/\Lambda$, $\Lambda<0$. Here there is a global timelike Killing vector corresponding to (z_0, z_4) rotation. It is $\bar{\xi}_a=(z_4,0,0,0,-z_0)$, $\bar{\xi}^2 = (z_\mu^2 + z_4^2)<0$, since $z_4 = z_0 = 0$ is excluded. There is one peculiar feature of anti-De Sitter space which should be mentioned. Specification of initial data on a complete spacelike surface does not lead to a unique prediction of the future state of a system (including gravity itself). Radiation not specified by the initial conditions can propagate in from infinity at a later time. Unlike the usual case where initial boundary conditions exclude incoming radiation thereafter, one is "safe" here only within evermore restricted regions of space at later times. Therefore, although the initial energy is perfectly well defined by the initial data, one can only extend the integration volume to all space at a later time if further (timelike) boundary conditions at infinity are imposed. It is in this sense that our stability results are to be understood, although the proof that energy is positive holds formally on any complete initial surface with no incoming radiation.

4. SMALL EXCITATIONS

We now apply the Killing energy together with the Killing vectors defined in the last two sections to discuss small oscillations about De Sitter or anti-De Sitter backgrounds. FOr this purpose, a canonical approach (Arnowitt, Deser & Misner 1959, 1960, 1962) is most useful to discuss the $O(h^2)$ part of $E(\xi)$. Indeed $T^0{}_\nu$ was derived canonically to this order long ago by Nariai and Kimura (1962). We will skip all details here, only noting that the excitations can be parametrized by the h_{ij} and their conjugate momenta p^{ij}, both being transverse traceless with respect to $\bar{g}$.

For $\Lambda>0$, the Hamiltonian density can be cast into the form

$$- T^o{}_o = \frac{1}{2} (f^{-3}(P^{ij})^2 + f(\nabla Q^{ij})^2)$$
$$- T^o{}_i = P^{ik}\partial_i Q_{jk} \qquad (4.1)$$

exactly as for the scalar field of (3.2), in terms of a canonically transformed set (Q,P). Not surprisingly, all the other features of the scalar model follow as well: $\int d^3x T^{0\nu}\bar{\xi}_\nu$ is conserved and is positive within the event horizon $\bar{\xi}^2<0$. However outside, when $\xi^2>0$, the energy is no longer positive, because the triangle inequality no longer applies, as with (3.5). One would expect all physical systems to behave in this way: the free part of the energy is always $\sim 1/2\int\{\pi^2+(\nabla\phi)^2\}$ while the momentum density is $\sim\pi\nabla\phi$. For physical matter, the non-linear parts of the energy are, like $V(\phi)$ in the scalar case, positive so the critical condition arises primarily at the free field level, where excitations beyond the horizon can give negative contributions. In particular, if the higher terms in $T^0{}_o(h)$ are effectively positive (as in the $\Lambda=0$ case), then the only De Sitter instability would be that due to the horizon.

For $\Lambda<0$, the small excitations are straightforwardly treated; this time only $T^0{}_o$ is required, since we can pick a coordinate system which is static and in which $\bar{\xi}_i=0$, and there is no horizon. The energy density is positive,

$$T^o{}_o \sim [p^2_{ij} + \frac{1}{4} (\nabla h_{ij})^2 + |\frac{\Lambda}{6}| h^2_{ij}] , \qquad (4.2)$$

and the system is stable. The masslike term in (4.2) is an artifact just like that of the spin 3/2 field (Deser & Zumino 1977) in cosmological supergravity; the gravitons have only two degrees of freedom since h_{ij} is transverse-traceless.

5. SEMI-CLASSICAL STABILITY FOR $\Lambda > 0$

Having shown that $\Lambda > 0$ solutions are stable to small fluctuations about the vacuum, at least within the horizon, one may make the further test of semi-classical stability in this case, i.e., look for Euclidean "bounce" solutions whose presence would signal quantum tunnelling instability. Of course even better would be proof that the total energy is positive, which will be given for $\Lambda<0$ in the next section; lacking this for $\Lambda>0$, we make some comments on bounces there. In general, topological effects can give stability problems in gravity (Brill & Deser 1973) and semi-classical instability has been found in other gravitational contexts (Perry 1981; Gross, Perry & Yaffe 1981; Witten 1981b).

A bounce solution here would be a metric which is asymptotically De Sitter and solves the Euclideanized Einstein equations. For example, the Euclidean continuation of the Schwarzschild-De Sitter metric would be a candidate. However, it is impossible to remove both the Schwarzschild and horizon singularities of this metric by the usual periodicity trick (Gibbons & Perry 1978). In terms of the Hawking picture, De Sitter space contains radiation at a temperature fixed by the value of Λ. If a black hole could form in this space with an intrinsic temperature less than this, it would grow forever by accretion. However, for the Schwarzschild-De Sitter black hole, the black hole temperature is always larger (Gibbons & Hawking 1977) than that of the exterior and the space is stable against this catastrophe.

Although it is doubtful on general grounds that bounce solutions exist for $\Lambda > 0$, it would clearly be desirable to extend the proofs of their absence (Witten 1981; Schoen & Yau 1979a) for $\Lambda = 0$ to this domain as well.

6. STABILITY FOR $\Lambda < 0$

We now show the Killing energy is positive for all excitations about the anti-De Sitter vacuum which vanish at infinity, and thereby establish stability in the $\Lambda < 0$ sector. We have already noted in Section 2 that all the generators of the graded anti-De Sitter algebra in supergravity are expressed as flux integrals, with the result that they obey the global algebra relations, in particular that of Eq. (2.19):

$$\{Q_{(\beta)},\bar{Q}_{(\beta')}\} = \frac{1}{2}\gamma^{(\mu)}_{(\beta\beta')}J_{(\mu 4)} + \sigma^{(\mu\nu)}_{(\beta\beta')}J_{(\mu\nu)}$$

We emphasize that in this expression, all indices are labels of particular Killing vectors or spinors. The explicit relations between the two are quite analogous to those holding in the Poincare case, and indeed can be essentially reduced to it because the $\tilde{D}_\mu$ commute; there exists (Gursey & Lee 1963) a transformation $\alpha = S\eta$ which reduces the equations $\tilde{D}_\mu\alpha = 0$ to $\partial_\mu\eta = 0$. A basis for the latter is given by, e.g., $\eta_\beta = \delta_\beta{}^{(\beta')}$. In any case, we may now simply treat the spinor "labels" (β), (β') in (2.19), which refer to the particular Killing spinor defining the corresponding charge $Q_{(\beta)}$, as normal flat space spinor indices. Multiplying (2.19) by the numerical matrix $\gamma^{(o)}_{\beta\beta'}$ and tracing gives the positivity relation for the operator $J_{(04)}$:

$$J_{(04)} = \sum_{\beta=1}^{4} Q_{(\beta)}Q_{(\beta)} \geqslant 0 \qquad (6.1)$$

since the $Q_{(\beta)}$ are real Majorana spinor operators. Now we just proceed as in the $\Lambda = 0$ case (Deser & Teitelboim 1977; Grisaru 1978), taking matrix elements of (6.1) with no on-shell fermions and go to the tree limit, $\hbar\to 0$. This implies that $E(\bar{\xi})$, which is just this limit of $J_{(04)}$, is positive for classical $\Lambda < 0$ gravity.

We also believe, although we have not carried out the details, that the recent purely classical proof of Witten (1981a) that energy is positive for $\Lambda = 0$ gravity can also be applied here. His proof, inspired by the supergravity argument, is based on considering solutions of the Dirac equation $\not{D}\varepsilon \equiv \gamma^i D_i\varepsilon = 0$ in an external metric satisfying $G_{o\mu} = 0$. From the relations

$$0 = \varepsilon^*\not{D}^2\varepsilon \equiv \varepsilon^*(D^2 + G_{o\mu}\gamma^o\gamma^\mu)\varepsilon = \varepsilon^* D^2\varepsilon \ ,$$

it follows upon integration that

$$\oint dS_i\,\varepsilon^* D^i\varepsilon = \int |\nabla\varepsilon|^2 d^3x \geqslant 0 \ . \qquad (6.2)$$

The surface integral is then separately shown to be proportional to E, which establishes positivity of the latter. The same reasoning should apply here with D_i replaced by $\tilde{D}_i$, and the metric now satisfying $G_{0\mu}+\Lambda g_{0\mu}=0$ provided, as is likely, the surface integral is again proportional to E. Similarly, it would be of interest to generalize the classical geometrical proof of Schoen and Yau (1979b, 1979c) to the $\Lambda < 0$ case. It may even be possible to establish full non-linear stability in the $\Lambda > 0$ case for excitations lying within the horizon by analytic continuation from $\Lambda < 0$, using the static form of the O(4,1) metric which covers the interior region only.

7. CONCLUSION

Our first task was to establish a unified and physically appropriate definition of energy in general relativity independent of the value of Λ. This was possible, for every Λ and for each set of solutions which tend asymptotically to a common background solution equipped with a timelike Killing vector. The associated energy is conserved, manifestly background gauge invariant and expressible as a flux integral at spatial infinity. We next studied stability by considering the positivity properties of the energy. For $\Lambda=0$ it was of course known that it is always positive for arbitrary asymptotically flat excitations. This property turned out to be shared by the $\Lambda<0$ models, as was demonstrated by a parallel study of the embedding supergravity theory. There, spinorial charges are also flux integrals and obey the anti-De Sitter graded global algebra, which easily implies positivity. For $\Lambda>0$, there is no grading possible and we studied several different stages. The first was that of small oscillations about the De Sitter background, whose energy was positive within the intrinsic event horizon. Those outside could exhibit a negative character associated with the transition of the Killing vector from timelike to spacelike at the horizon. This phenomenon was seen to be entirely independent of the system and is a simple consequence of the form of any free-field Hamiltonian; it marks the onset of Hawking radiation. Apart from this property, we found no evidence for instability at the semi-classical level, and indeed we believe that energy inside the horizon is positive for general excitations there. The resemblance between these properties of gravity models, irrespective of whether $\Lambda=0$, provides a unified picture, but also implies that $\Lambda\neq 0$ theory cannot be excluded on stability grounds alone.

ACKNOWLEDGEMENTS

This research was supported in part by NSF grant PHY7809644A02.

REFERENCES

L.F. Abbott and S. Deser (1982), Nucl. Phys. B195, 76.

R. Arnowitt, S. Deser and C. W. Misner (1959), Phys. Rev. 116, 1322; (1960) 117, 1595 and in (1962), Gravitation: an introduction to current research, ed. L. Witten (Wiley, New York).

D. Brill and S. Deser (1968), Ann. Phys. 50, 548.

D. Brill and S. Deser (1973), Comm. Math. Phys. 32, 291.

S. Coleman (1977), Phys. Rev. D15, 2929.

S. Coleman and C. G. Callan (1977), Phys. Rev. D16, 1762.

S. Deser and C. Teitelboim (1977), Phys. Rev. Lett. 39, 249.

S. Deser and B. Zumino (1977), Phys. Rev. Lett. 38, 1433.

G. W. Gibbons and S. W. Hawking (1977), Phys. Rev. D10, 2738.

G. W. Gibbons and M. J. Perry (1978), Proc. Roy. Soc. A358, 467.

M. Grisaru (1978), Phys. Lett. 73B, 207.

D. Gross, M. J. Perry and L. Yaffe (1981), Princeton University preprint.

F. Gursey and T. D. Lee (1963), Proc. Nat. Acad. Sci. 49, 179.

S. W. Hawking and G. F. R. Ellis (1973), The Large Scale Structure of Space-time, (Cambridge).

H. Nariai and T. Kimura (1962), Progr. Theor. Phys. 28, 529.

M. J. Perry in (1981) Superspace and Supergravity, eds. S. W. Hawking and M. Rocek (Cambridge).

P. Schoen and S. T. Yau (1979a), Phy. Rev. Lett. 42, 547; (1979b), Comm. Math. Phys. 65, 45; (1979c) Phys. Rev. Lett. 43, 1457.

P. K. Townsend (1977), Phys. Rev. D15, 2802.

E. Witten (1981a), Comm. Math. Phys. 80, 381; (1981b), Princeton University preprint.

QUANTUM GRAVITY AT SHORT DISTANCES

G. Furlan*

Istituto di Fisica Teorica, Università di Trieste

International Centre for Theoretical Physics, Trieste

* Supported in part by I.N.F.N., Sezione di Trieste

The recent progress in Particle Physics, in particular the unified theories of weak, electromagnetic and (perhaps) strong interactions, are getting us accostumed to a new role of the boundary conditions which are becoming an essential part of the fundamental laws of physics, rather than a distinct element. In other words the Lagrangian one starts with, often represents only one of the ingredients of the complete description of a given physical phenomenon and it has to be supplemented by a specific mechanism (the spontaneous breaking of a symmetry) which allows the introduction in the theory of a dimensional scale. The application of these ideas to gravitation as a quantum field theory has been proposed by several authors [1]. The common aim is to clarify the role of the Newton's constant $G_N \simeq 10^{-38}\,\mathrm{GeV}^{-2}$; this dimensional quantity determines on the macroscopic scale the entity of the gravitational phenomena but leads to a non renormalizable quantum field theory. Almost all the attempts are based on a, sometime drastic, modification of the original Einstein-Hilbert lagrangian: quantum gravity is considered as a spontanously broken theory (this requires the presence of an additional scalar field) or as the manifestation of dynamical symmetry breaking (induced gravity) up to the versions of a composite model for the graviton [2]. The present form of gravity would just appear as a low energy manifestation of a more fundamental (and hopefully satisfactory) theory.

A more conventional, so to say, proposal has been recently put forward by de Alfaro, Fubini and Furlan [3] and I want to summarize its main points.

1. The starting point is represented by an analysis of the correct value of the "scale dimension" one has to ascribe to the metric tensor $g_{\mu\nu}$ as a (quantum) field.

From the transformation property under an infinitesimal re-parametrization of coordinates, $x'^{\mu} \simeq x^{\mu} - \varepsilon^{\mu}(x)$,

$$\delta g_{\mu\nu} \equiv g'_{\mu\nu}(x) - g_{\mu\nu}(x) = \varepsilon^{\lambda}\partial_{\lambda} g_{\mu\nu} + g_{\lambda\nu}\partial_{\mu}\varepsilon^{\lambda} + g_{\mu\lambda}\partial_{\nu}\varepsilon^{\lambda} , \qquad (1)$$

one realizes that in the particular case of a dilatation $\varepsilon^{\mu}(x) = -\varepsilon x^{\mu}$

$$\delta g_{\mu\nu} = \varepsilon(x \cdot \partial + 2) g_{\mu\nu} \qquad (2)$$

This suggests, quite naturally, that $g_{\mu\nu}(x)$ has dimension $\Delta = -2$ (in units of length); as a consequence its inverse $g^{\mu\nu}(x)$, $\equiv \det g_{\mu\nu}(x)$, $R = g^{\mu\nu} R_{\mu\nu}$ have $\Delta = 2, -8, 0$ and so on.

In general the value of Δ is determined by the tensor nature of the field and it is easy to recognize that Δ = (number of controvariant) - (number of covariant components).

It then follows that the Einstein-Hilbert form of the action, namely

$$A_E = -\frac{1}{4} \int d^4x \sqrt{-g}\, R, \qquad (3)$$

is automatically dilatation invariant (as it must, given the invariance of the theory under general coordinate transformations) with no room for dimensional parameters. This was the case in the conventional formulation where $1/G_N$ actually multiplies the integral of eq. (3). (On pure invariance grounds at least an additional term is allowed i.e. $\lambda\sqrt{-g}$ where λ is the so called cosmological constant).

2. Comparison with the classical formulation where $g_{\mu\nu}(x)$ is taken to be dimensionless suggests the simple way the Newton's constant can be let to enter the theory.

At the purely classical level G_N appears through the elementary solution to the equations of motion in empty space ($R_{\mu\nu} = 0$) represented by the flat tensor:

$$g_{\mu\nu}(x)\,|_{\text{flat}} = \frac{1}{4\pi G_N}\,\eta_{\mu\nu} \tag{4}$$

According to this point of view the dimensional Newton's constant does not represent a general property of the gravitational action but it is rather related to the class of solutions which characterize the large distance behaviour of the theory. It is clear that classically such a formulation is completely equivalent to the conventional one and we can pass from one to the other by just redefining the field ($g_{\mu\nu} \to G_N^{-1/2}\, g_{\mu\nu}$).

(However when the coupling with matter fields is considered some new features, independent of G_N, can appear even for the classical solutions) [4].

The quantum language reveals the actual meaning of a formulation of gravity based on the action given in eq. (3). In order to regain the Newton constant we have to interpret it as a vacuum effect namely

$$\langle 0 \,|\, g_{\mu\nu}(x) \,|\, 0 \rangle = \frac{1}{4\pi G_N}\,\eta_{\mu\nu} \tag{5}$$

Having such a constant, non vanishing, vacuum expectation value of the field $g_{\mu\nu}(x)$ is the signal of spontaneous breaking of dilatation invariance. We are thus facing the by now familiar phenomenon of a theory whose starting formulation exhibits invariance properties (in this case under general coordinate transformations) that the vacuum has not. This lack of invariance of the vacuum is then the way the dimensional constant G_N slips in.

3. The previous discussion should have made evident the purpose of exploiting the lesson from the study of massive gauge theories in order to achieve an improved understanding of the renormalization problem of quantum gravity. In particular the fact that "the divergence structure of a field theory respects the symmetry of the Lagrangian even if the vacuum does not" [5], which is at the root of the successful renormalizability of gauge theories, strongly suggests the usefulness of an analogous line of investigation in our case.

One can thus expect that some general quantum features of the Green functions, like the small distance behaviour should be almost independent of G_N and substantially be fixed by the invariance properties of the underlying action, in particular dilatation invariance. If so, quantum gravity should not behave differently from a renormalizable theory.

The practical realization of such a programme is hampered by the non polynomial character of the gravitational lagrangian, which contains inverse operators and all that. This forces a treatment where the field $g_{\mu\nu}(x)$ is given by a classical background plus a quantum part in which one expands.

However it is possible to show on some selected examples that dimensional arguments are still of value. For simplicity consider the case of a scalar field of zero mass and dimension -1, which we write as a classical plus a quantum piece:

$$\phi(x) = \mu + \phi'(x), \quad \langle \phi'(x) \rangle_0 = 0 \tag{6}$$

The (euclidean) Green function has the form

$$\langle \phi(x) \ \phi(y) \rangle = \mu^2 + \frac{1}{4\pi^2 (x-y)^2} \tag{7}$$

and the small distance behaviour $\sim \frac{1}{(x-y)^2}$ is the one required by naive dilatation arguments.

It is interesting that the same indication can be shown to persist for less trivial, non polynomial, objects. For instance for the quantity $\langle \phi^{-1}(x) \ \phi^{-1}(y) \rangle_0$ one finds, after a careful resummation of all terms of the $\phi'(x)/\mu$ series, that

$$\langle \phi^{-1}(x) \ \phi^{-1}(y) \rangle_0 \underset{(x-y) \to 0}{\sim} (x-y)^2 \ \ell n \, [\mu^2 (x-y)^2] \tag{8}$$

The ultraviolet behaviour is the one following from simply counting the dimensions, apart from the logarithmic term which embodies the residual (weak) dependence on the background.

This argument can be generalized to more complicated functions and more interesting theories. Thus for gravitation one has

$$\langle g(x)^{-1/2} \ g(y)^{-1/2} \rangle_0 \sim [(x-y)^2]^4 \ \ell n \, [\mu^2 (x-y)^2] \tag{9}$$

and similar ones.

4. For a general discussion of gravitation it seems however that the only practical procedure presently available has to rely on an expansion of some sort. It is customary to use the Newtonian solution(4) as classical background and write

$$g_{\mu\nu}(x) = \frac{1}{4\pi G_N} \left(\eta_{\mu\nu} + G_N^{1/2} \phi_{\mu\nu} \right) \quad (10)$$

with $\phi_{\mu\nu}$ an operator of dimension -1. The standard procedure is then to expans in $G_N^{1/2} \phi_{\mu\nu}$: due to the dimensionality of the Newton's constant all the troubles of a non renormalizable theory show up.

The blame has to be put on the separation (10) which breaks general invariance and therefore dilatation invariance. As a consequence by keeping only a finite number of terms, which individually are more and more singular, one will never be able to reproduce the properties of the Green functions, for instance the small distance behaviour, which are predicted to follow from scale invariance arguments.

Thus the expansion given in eq. (10), which is no doubt adequate in the infrared domain, is not suited for a perturbative-like discussion of the theory at small distances. If dilatation invariance predictions have to be tested a decomposition for $g_{\mu\nu}(x)$ is required, which is preserved by those transformations.

A possibility can be to write, for the sake of investigation,

$$e_\mu^a(x) = h(x)\, \delta_{\mu a} + \bar{e}_\mu^a(x) \quad (11)$$

where $e_\mu^a(x)$ is the tetrad field, introduced for convenience, and

$h(x)$ and $\bar{e}_\mu^{\ a}(x)$ are both operators of dimension -1 [6]. The separation is scale invariant and the expansion, grossly speaking, is in $\bar{e}/h$, an object of dimension zero. It turns out that the "interaction" part of the gravitational Lagrangian is of the form

$$L_I = \partial_{\mu_1} e^a_{\nu_1} \partial_{\mu_2} e^b_{\nu_2} I^{ab}_{\mu_1\nu_1\mu_2\nu_2} (\bar{e}/h) \qquad (12)$$

and has therefore dimension -4 at any order in $\bar{e}/h$. Apart from first order logarithms, of the kind present in eq. (8) or (9), such a model should not be too drastically different in same respects, from a renormalizable case.

Even if heuristic the example is instructive : it confirms the suggestion that the uniquely rich structure of Einstein's theory can represent a still unexploited clue to the problem of Quantum Gravity [7].

References

[1] A.D. Sakharov, Dokl. Akad, Nauk, SSR 177 (1967) 70; (Soviet Physics Dokl. 12 (1968) 1040);
Ya.B. Zel'dovich, Zh.ETF Pis. Red. 6 (1967) 922; (JETP Lett. 6 (1967) 345);
O. Klein, Phys. Scr. 9 (1974) 69.
P. Minkowski Phys. Lett. 71B (1977) 619;
K. Akama, Y. Chikashige, T. Matsuki and H. Terazawa, Progr. Theor. Phys. 60 (1978) 868;
K. Akama, Progr. Theor. Phys. 60 (1978) 1900;
L. Smolin: Nucl. Phys. B160 (1979) 253;
S.L. Adler, Phys. Rev. Lett. 44 (1980), Phys. Lett. 95B (1980) 241;
B. Asslacher and E. Mottola, Phys. Lett. 95B (1980) 237;
A. Zee, Phys. Rev. D23 (1981) 858.

[2] D. Amati and G. Veneziano: "Metric from matter" CERN preprint, Ref. TH 3126 (1981).
[3] V. de Alfaro, S. Fubini and G. Furlan: Il Nuovo Cimento A50, 523 (1979) and Il Nuovo Cimento B57, 227 (1980).
[4] Several aspects of classical solutions are reviewed in V. de Alfaro, S. Fubini and G. Furlan, Acta Physica Austriaca Sup. XXII 51, (1980).
[5] S. Coleman in "Laws of Hadronic Matter" ed. A. Zichichi, (Acad. Press, N.Y. 1975).
[6] V.de Alfaro, S. Fubini and G. Frulan: Phys. Letters 97B, 67 (1980).
[7] A.Salam: "Impact of quantum gravity in Particle Physics" in "Quantum Gravity, an Oxford Symposium" eds. C. Isham, R. Penrose, D. Sciama, (Oxford Univ. Press, 1975).

QUANTUM THEORY OF WORMHOLES

P. Hajicek

Institute for Theoretical Physics

Sidlerstrasse 5

CH-3012 Bern, Switzerland

Abstract: We generalize the quantum theory of solitons to the wormholes in the Einstein-Maxwell model. The main difficulty, the singularity of the fields at r = 0, is circumvented by cutting away the unphysical part of spacetime and imposing boundary conditions at the resulting boundary. Semiclassical formula (sum of all three graphs) is obtained for the transition amplitude of relativistic processes with a single wormhole. The generating functional for Green's functions in one soliton sector is determined to the zero approximation; suitable gauges are found which simplify its calculation. Finally, some scattering processes of photons and gravitons off the wormholes are studied.

1. Introduction

In his book [1], Wheeler estimated the vacuum fluctuations of the Einstein-Maxwell fields and found that small wormholes of about one Planck mass and one electronic charge could form.

The Einstein-Maxwell equations have stationary classical solutions of the wormhole type: the Kerr-Newman family. This is a four-parameter set of electrovacuum fields, the parameters being m, a, e, and b, where m/G is the mass, am/G the angular momentum, $e/\sqrt{G}$ the electric, $B/\sqrt{G}$ the magnetic charge of the hole, and G the Newton constant (we use the units with $\hbar = c = 1$). Several authors have speculated that these solutions could play the role of solitons in the quantum gravity [2, 3, 4].

An interesting discovery in the quantum theory of solitons is due to Coleman [5]. Two different field-theoretical models with solitons can be in the relation of duality to each other: ordinary field excitations in one model become solitons or compound particles of the other model. An example is the so-called sine-Gordon system and the massive Thirring system. In fact, first the perturbation theory of the latter system yields the full-fledged quantum theory of the solitons in the former.

If the Kerr-Newman solutions really are solitons, then there could be a dual theory to the Einstein-Maxwell one. Such a dual theory would provide the proper mathematical apparatus to describe the wormhole-like excitations and would, therefore, enable a detailed development of Wheeler's idea.

These and similar [6] speculations motivate the effort aimed at a generalization of the current soliton theory (for reviews, see, e.g. [7, 8, 9, 10]) to the wormholes. The main difficulty is the singularity of the wormhole solutions at r = 0. We propose to cut away the unphysical singular region of space-time, and to define the dynamics of fields and wormholes by suitable conditions on the fields at the resulting boundary.

The corresponding quantum theory, which is based on a perturbation expansion, could be developed for one soliton sector. Here, we are able to sum all three graphs and to write down the formula for the generating functional for Green's functions in the zero approximation. In this way, a basis is layed down for the speculations in [4] and [6]. We have also estimated the cross section for capture of gravitons and photons on the wormhole, which is an unusual effect in the soliton physics.

In the present paper, we give a review of the results and methods. The detail of calculations and proofs will be published elsewhere.

2. Boundary conditions

We describe the system in the way due to Arnowitt et al. [11]. The notation is taken over from [11]. At the flat infinity, the fields satisfy the following boundary conditions

$$(1)\quad \begin{cases} N = 1 + O(r^{-1}) \,, \\ N_k = -\dot{X}^k + O(r^{-1}) \,, \\ g_{kl} = \delta_{kl} + O(r^{-1}) \,, \end{cases}$$

$$(2)\quad \begin{cases} A_\mu = O(r^{-1}) \,, \\ F_{\mu\nu} = O(r^{-2}) \,, \end{cases}$$

allowing for the shift of the system by an arbitrary time dependent vector $X^k(t)$.

At the wormhole boundary B, which is defined by the equation $x^1 = x^1_B$, our conditions are as follows:

$$(3) \qquad N\Big|_{x^1 = x^1_B} = 0 \,, \qquad N_k n^k\Big|_{x^1 = x^1_B} = 0 \,.$$

The values of the following quantities

$$(4) \qquad g_{KL}\Big|_{x^1 = x^1_B} \,, \qquad A_K\Big|_{x^1 = x^1_B} \,,$$

$$(5) \qquad \left(P^k{}_m \frac{\pi^{m\ell}}{\sqrt{g}} n_\ell\right)_{x^1 = x^1_B} \,, \qquad \left(\frac{\mathcal{E}^k}{\sqrt{g}} n_k\right)_{x^1 = x^1_B} \,,$$

are all given and time independent. Here, n^k is the unite normal vector to B, $P^k{}_m$ is the projection operator to the tangent space of B, capital indexes run through 2, 3, and the meaning of other quantities is as in [11].

One of the roles of the boundary conditions (3) - (5) is to determine the soliton solution uniquely. Notice that there are many wormhole solutions to Einstein-Maxwell equations: e.g. the whole Kerr-Newman family with its four parameters m, a, e, and b. Thus, our boundary conditions must distinguish between different members of this family. We have the following theorem:

Theorem 1: Let D be the domain of outer communication of a space-time from the Kerr-Newman family. Let the corresponding values of the quantities (4) and (5) be given. Then, the parameters m, a, e and b of D are uniquely determined.

The physical meaning and transformation properties of the quantities (4) and (5) can be given as follows. There are some basic fields describing the geometrical structure of any Killing horizon, called in [12]: γ_{KL}, Ω^K, E and H. We have:

$$(6) \qquad \gamma_{KL} = g_{KL}\Big|_B \,, \qquad \Omega^K = \left(P^K{}_r \frac{\pi^{rs}}{\sqrt{g}} n_s\right)_B \,,$$

$$(7) \qquad E = \left(\frac{\mathcal{E}^r}{\sqrt{g}} n_r\right)_B \,; \qquad H = \left[\frac{1}{\sqrt{\gamma}}\left(\partial_2 A_3 - \partial_3 A_2\right)\right]_B \,,$$

where $\gamma = \mathrm{Det}(g_{KL})$.

Finally, one can check that the quantities (4) and (5) have vanishing Poisson brackets with each other.

In a suitable gauge, the boundary conditions (3) - (5) are equivalent to the conditions (1) - (6) in [4].

3. The action

In this section, we will find the form of the action functional for the system defined by our boundary conditions, and separate from it those terms which govern the motion of the system as a whole.

The action I in the original ADM form [11] is convenient for spacetimes with compact space slices. If we calculate the variation of I, not discarding divergences as usual, we find, in addition to the usual terms, the term $(\delta I)_s$ of the form

$$(\delta I)_s = \frac{1}{16\pi G} \int_{t_1}^{t_2} dt \oint_{\partial S(t)} ds_\ell \, v^\ell \quad ,$$

where

$$(8) \quad v^\ell = G^{rsk\ell} (N \nabla_k \delta g_{rs} - N_{,k} \delta g_{rs}) + 2 N_k \delta\pi^{k\ell} +$$
$$+ (N^r \pi^{s\ell} + N^s \pi^{r\ell} - N^\ell \pi^{rs}) \delta g_{rs} + N \frac{1}{\sqrt{g}} \epsilon^{\ell rs} B_r \delta A_s +$$
$$+ (N^\ell \mathcal{E}^r - N^r \mathcal{E}^\ell) \delta A_r - A_0 \delta \mathcal{E}^\ell \quad ,$$

and

$$G^{rsk\ell} = \frac{1}{2}\sqrt{g}\,(g^{rk} g^{s\ell} + g^{s\ell} g^{r\ell} - 2 g^{k\ell} g^{rs}) \quad .$$

We have, therefore, to add a surface term, I_s , to the ADM action such that $\delta I_s = -(\delta I)_s$. Even if the form of δI_s is prescibed, the form of I_s depends on the boundary conditions and, for some boundary conditions, it need not exist at all. Let us find I_s at B. Using (3), we find after some algebra:

$$(9) \quad \delta I_S^B = \int_{t_1}^{t_2} dt \, \frac{1}{16\pi G} \oint_B dx^2 dx^3 \Big[2\varkappa\sqrt{\gamma} + 2N^K \delta(\sqrt{\gamma}\,\Omega_K) + A_0\, \delta(\sqrt{\gamma}\,E) - \mathcal{E}^1 N^K \delta A_K \Big] .$$

Here, $\varkappa$ is defined by $N_{,k} = -\varkappa n_k$. (4) - (7) lead immediately to the choice: $I_S = 0$. Thus, no surface terms at B have to be added to I.

At the flat infinity, the surface term I_S^∞ has been found in [13]. Using this result, we obtain the Lagrangian for our system in the form

$$(10) \quad L = \frac{1}{16\pi G} \int_{S(t)} d^3x \left(\pi^{kl} \dot{g}_{kl} - \mathcal{E}^k \dot{A}_k - \lambda^\alpha \mathcal{T}_\alpha \right) - E + N^k_\infty \Pi_k ,$$

where the electromagnetic quantities are rescaled by $2\sqrt{G}$, λ^α , α = 0, 1, 2, 3, 4, are the Lagrange multipliers N, N_1, N_2 , N_3 and A_0, respectively, and $\mathcal{T}_\alpha$ are the corresponding constraints (see [11]). Further:

$$N^k_\infty = \lim_{r\to\infty} N^k ,$$

$$(11) \quad \Pi_k = -\frac{1}{8\pi G} \oint_{r=\infty} ds_l \, \pi_k{}^l ,$$

and

$$(12) \quad E = \frac{1}{16\pi G} \oint_{r=\infty} ds_l \left(g_{kl,k} - g_{kk,l} \right) ,$$

see [13]). Π_k Is the total momentum and E the total energy of the system.

Our next task is to write the Lagrangian in such a way that the total momentum, P_k , and the center of mass coordinate, X^k , appear explicitly in it. An elegant method is described in [13]. We have the constraints

$$p_k \equiv \Pi_k - P_k = 0 ,$$

and the extended Lagrangian reads

$$(13)\quad L_E = \frac{1}{16\pi G}\int_{S(t)} d^3x\left(\pi^{kl}\dot{g}_{kl} - \mathcal{E}^l\dot{A}_l - \lambda^\alpha \mathcal{J}_\alpha\right) + P_k\dot{X}^k - E + N^k_\infty p_k \,.$$

L_E has more gauge invariance then L:

Theorem 2: Let $Y^k(t,x^l)$ be a smooth vector field which is tangential to S(t) for all t and satisfies:

$$(14)\qquad Y^k(t,x^l) - Y^k(t) = O(r^{-1})$$

at $r\to\infty$, where $Y^k(t)$ is a triple of time functions and

$$(15)\qquad Y^k(t,x^l)\big|_B = 0 \,.$$

Otherwise, $Y^k(t,x^l)$ is arbitrary. Then, L_E is invariant with respect to the transformation

$$(16)\quad \begin{cases} \tilde{x}^k = x^k - Y^k(t,x^l)\,, \quad \tilde{t} = t\,, \\ \tilde{X}^k = X^k + Y^k(t)\,, \\ g_{\mu\nu}(x) = \dfrac{\partial\tilde{x}^\rho}{\partial x^\mu}\dfrac{\partial\tilde{x}^\sigma}{\partial x^\nu}\tilde{g}_{\rho\sigma}(\tilde{x})\,, \quad A_\mu(x) = \dfrac{\partial\tilde{x}^\rho}{\partial x^\mu}\tilde{A}_\rho(\tilde{x})\,. \end{cases}$$

This gauge invariance is analogous to that found for scalar field action in two dimensions [14], if one extends the Lagrangian by terms analogous to (13).

In the quantum theory, we have to remove gauge freedom by gauge conditions. In addition to the usual gauge conditions

$$(17)\qquad c^\alpha(x) = 0\,,$$

we shall have three new ones:

$$(18)\qquad C^k = 0\,,$$

which suppress the freedom (16). We observe from (16) that neither $c^\alpha(x)$ nor C^k have to depend on X^k. Let us assume, they are chosen in this way.

4. The soliton solution

As examples of wormholes, we consider the spherically symmetric, static, Reissner-Nordström family with parameters m, e and b. The metric and the potential are usually given by

$$ds^2 = -\Delta dt^2 + \frac{1}{\Delta} dr^2 + r^2 d\theta^2 + r^2 \sin^2\theta d\varphi^2, \tag{19}$$

$$A_\mu dx^\mu = -\frac{e}{r} dt - b\cos\theta d\varphi, \tag{20}$$

where

$$\Delta(r) = \frac{r^2 - 2mr + e^2 + b^2}{r^2}. \tag{21}$$

At B, the coordinate r is not regular. We can use x^1, defined by

$$x^1 = \int_{2m}^{r} \frac{d\xi}{\sqrt{\Delta(\xi)}} :$$

$$x^1_B = \frac{1}{2} m \lg \frac{m - \sqrt{e^2+b^2}}{m + \sqrt{e^2+b^2}} - \sqrt{e^2+b^2},$$

which is $-\infty$ for extreme case. Then, at B, the solution in these coordinates fulfills all boundary conditions (3) - (5). At the infinity $r = \infty$, the coordinates r, θ, φ are not convenient. We perform the transformation

$$\bar{x}^1 = (r-m)\sin\theta\cos\varphi,$$

$$\bar{x}^2 = (r-m)\sin\theta\sin\varphi,$$

$$\bar{x}^3 = (r-m)\cos\theta.$$

The transformed metric, $\bar{g}_{\mu\nu}$, and potential, $\bar{A}_\mu$, satisfy (1) and (2). Another useful system of coordinates is x^μ defined by the transformation

$$\bar{x}^\mu = \Lambda^\mu{}_\nu (x^\nu + x^0 v^\nu),$$

where

$$v^\nu = (0, v^k) ,$$

$$\Lambda^\mu{}_\nu = \begin{pmatrix} \gamma & , -\gamma v^l \\ -\gamma v^k & , \delta_{kl} + \frac{\gamma^2}{1+\gamma} v^k v^l \end{pmatrix} ,$$

$$\gamma = \frac{1}{\sqrt{1-v^2}} .$$

In this system, the linear momentum Π^S_k, given by (11), and the energy E^S, given by (12), are

$$(22) \qquad \Pi^S_k = \frac{m}{G} \gamma v^k ,$$

$$(23) \qquad E^S = \frac{m}{G} \gamma .$$

We can use the coordinates x^0, x^1, x^2, x^3 near the infinity $r = \infty$ and t, r, θ, φ near B; the slicings represented by t and x^0 can be smoothly joined by some interpolating surfaces in such a way that the resulting metric and potential do not depend on the corresponding common time coordinate, which we shall denote by t. The solution prepared in this way will be denoted by $g^S_{\mu\nu}$, A^S_μ.

5. The formula for Z_0

We go over to the quantum theory in one-wormhole sector. The transition amplitude between an initial and a final state described by the wave functionals

$$\psi_{i,f}\left[g_{kl}(x), A_k(x), X^k\right]$$

is given by the following functional integral

$$\langle \psi_f | \psi_i \rangle = \int d\mu \, dP_k \, dX^l \, \psi_f^* \psi_i \, \delta(c^\alpha(x)) \, \delta(C^k) \, .$$

$$. \, \mathrm{Det}\left\{ c^\alpha(x), C^k ; J_\beta(y), p_l \right\} e^{i \int_{t_1}^{t_2} dt \, L_E} ,$$

where

$$d\mu = dg_{kl}\, d\pi^{rs}\, dA_m\, d\mathcal{E}^m\, dX^\alpha\, dN^k_\infty \ .$$

We can choose

$$\psi_{i,f} = e^{iP^{i,f}_k X^k}\, \tilde{\psi}_{i,f}[g_{kl}, A_k] \ .$$

As the gauges $c^\alpha(x)$ and C^k are independent of X^k, the integration over P_k and X^k is trivial and yields

$$(24) \quad \langle\psi_f|\psi_i\rangle = \delta^3(\vec{P}_f - \vec{P}_i)\int d\mu\, \tilde{\psi}^*_f\, \tilde{\psi}_i\, \delta(c^\alpha(x))\, \delta(C^k) \ .$$

$$\cdot\, \mathrm{Det}\,\{c^\alpha(x), C^k; \mathcal{J}_\beta(y), p'_l\}\cdot \exp i\int_{t_1}^{t_2} dt\,(L - N^k_\infty \Pi_k + N^k_\infty p'_k) \ ,$$

where p'_k is given by

$$(25) \qquad p'_k = \Pi_k - P^i_k \ .$$

Next, we shift all integration variables in (24) by their values for $g^s_{\mu\nu}$ and A^s_μ, rescaling the new variables by $\sqrt{G}$. We obtain

$$\int_{t_1}^{t_2} dt(L - N^k_\infty \Pi_k + N^k_\infty p'_k) = \int_{t_1}^{t_2} dt\, E^s + I_R \quad ,$$

where I_R abbreviates the rest on the R.H.S. With (23), (24) becomes

$$(26) \quad \langle\psi_f|\psi_i\rangle = e^{-i\frac{m\gamma}{G}(t_2 - t_1)}\, \delta^3(\vec{P}_f - \vec{P}_i) \ .$$

$$\cdot \int d\mu'\, \tilde{\psi}^*_f\, \tilde{\psi}_i\, \delta(c^\alpha(x))\, \delta(C^k)\, \mathrm{Det}\,\{c^\alpha(x), C^k; \mathcal{J}_\beta(y), p'_l\}\, e^{iI_R} \ .$$

The first two factors on the R.H.S. of (26) coincide with the transition amplitude $\langle P_f|P_i\rangle$ for a free relativistic particle of mass m/G. This is analogous to the results of the quantum theory of scalar field solitons in two spacetime dimensions [14] and gives exactly the sum of all tree graphs for single soliton processes.

The integral contains corrections to it, and can be formally calculated by the usual perturbation method [15]. The generating functional for Green's functions is obtained, if we add the following source term to the action

$$\Sigma = \int_{t_1}^{t_2} dt \int_{S(t)} d^3x \left(\mathcal{J}^{kl} \delta g_{kl} + K_{kl} \delta \pi^{kl} + j^k \delta A_k + k_l \delta \mathcal{E}^l \right),$$

rotate the time variable in the complex plane as usual, and go over to the limit $t_1 \to -\infty$, $t_2 \to +\infty$. The boundary of the Euclidean manifold obtained in this way consists of the surfaces $t = \pm\infty$, $r = \infty$, and $r = r_B$. The Dirichlet propagator (whose analytic continuation will play the role of Feynman propagator) for the linearized theory has to vanish on all of these surfaces. This is in agreement with the boundary conditions (1) - (5), and corresponds to the choice of Boulware's vacuum [16] as the zero approximation to the ground state in the one-soliton sector.

In this way, we obtain

$$(27) \quad Z(\mathcal{J}, K, j, k) = \int d\mu' \, \mathcal{D}et \{ c^\alpha(x), C^k; \mathcal{J}_\beta(y), p'_l \} \cdot \delta(c^\alpha(x)) \delta(C^k) \, e^{i(I_R + \Sigma)} .$$

If we choose $\vec{P}_i = O(0)$, then $g^S_{\mu\nu}$ and A^S_μ differ from $\bar{g}^S_{\mu\nu}$ and $\bar{A}^S_\mu$ by terms of order 1 or higher. These can be included in small disturbances, and the expansion of the action can start with $\bar{I}_R$: the action for small disturbances on the background $\bar{g}^S_{\mu\nu}$, $\bar{A}^S_\mu$. We want to calculate Z_0, that is to approximate (27) only by terms of the order lower or equal to zero. We obtain, to this order,

$$(28) \quad I_R + \Sigma = I_0 + \frac{1}{2} \delta N^k_\infty \delta \Pi_k + \Sigma ,$$

where I_0 is the second variation of the action on the static, spherically symmetric background.

I_0 has been studied in detail by Moncrief [17], [18]. He introduced new pairs of canonically conjugated variables such that the constraints $\mathcal{J}_\alpha(x)$ becomes either the new momenta in some pairs, or the new coordinates in other pairs. These variables are very useful in calculation of Z_0, because they discouple the field equations and enable to choose gauges $c^\alpha(x)$ and C^k such that the determinant in (27) becomes 1. Of course, we must guarantee that these gauges are compatible with the boundary conditions (1) - (5). This can be done as follows.

For odd parity disturbances with angular momentum number $L \geqslant 2$, Moncrief's variables are (k_1, π_1), (k_2, π_2) and (f_1, π_f) (see [17]). The only constraint is

$$\pi_2 = 0$$

and we choose the gauge to be

$$k_2 = 0 \ .$$

Then, the boundary conditions (4), (5) imply for Moncrief's radial functions $\mathcal{P}_\pm$ (see [19])

$$\mathcal{P}_\pm \big|_B = 0$$

and (3) follows.

For odd parity disturbances with L = 1, Moncrief's variables are (k_1, π_1) and (f, π_f) (see [18]). The constraint is

$$\pi_1 = C$$

where C is a constant. We choose the gauge

$$k_1 = 0 \ .$$

Then, (4) and (5) imply C = 0 and

$$\mathcal{P}_f \big|_B = 0 \ ,$$

where $\mathcal{P}_f$ is the radial function ([19]). Again, (3) follows.

For even parity disturbances with $L \geqslant 2$, Moncrief's variables are (q_1, π_1), (q_2, π_2), (q_3, π_3), (q_4, π_4), (F, π_f), (a_2, f_2) (see [18]), the constraints are

$$q_2 = \pi_3 = \pi_4 = f_2 = 0$$

and we choose the gauges

$$q_3 - \frac{2}{L(L+1)\Lambda r} q_1 = q_4 = \pi_2 = a_2 = 0 \ ,$$

where Λ is given in [18]. The conditions (4) and (5) imply for the radial functions $\mathcal{R}_\pm$:

$$\mathcal{R}_{\pm}\big|_B = 0$$

and (3) follows.

Finally, for the even parity disturbances with L = 1, we have the variables (q_1, π_1), (q_2, π_2), (q_3, π_3), (H, π_H) and (f_2, a_2). The constraints are

$$\pi_1 = q_2 = \pi_3 = f_2 = 0$$

and the gauges can be chosen as follows

$$q_1 = \pi_2 = q_3 = a_2 = 0 \ .$$

Then, (4) and (5) again imply (3) and

$$H\big|_B = 0$$

where H is the radial function (see [18]).

The gauge conditions C^k which makes the determinant in (27) equal 1 can be found as follows. By transforming the variables, Moncrief discarded surface terms. This is not allowed with our boundary conditions at $r=\infty$. We find that a surface term of the form

$$16\pi G \left(-\tfrac{2}{3}\,\delta\Pi_k\, \delta\dot{X}^k_\infty\right)$$

remains in the action, and we can always choose coordinates such that

$$\delta X^1_\infty = \delta X^2_\infty = 0 \ ,$$

$$\delta X^3_\infty = -\frac{1}{4}\sqrt{\frac{3}{4\pi}}\left(\frac{r^2}{4mr - 3e^2 - 3b^2}\, q_1 - r q_3\right)_{r\to\infty} \ ,$$

q_1 and q_2 are Moncrief's variables for even parity, L = 1, disturbances. Hence

$$C^k = -\frac{2}{3}\,\delta X^k \ .$$

Another important question is the problem of zero modes. For L = 0, there are three such modes: two even parity ones, corresponding to a variation δm and δe of the parameters of the family, and one odd parity one, corresponding to δb . For L = 1, there are three odd parity zero modes, corresponding to a small net angular momentum of the hole, that is to δa and a direction of rotation axis. All these six modes violate the boundary conditions (3) - (5). They are, therefore, excluded, and we can construct a regular propagator.

6. High energy capture in the case $e^2 + b^2 = m^2$

It is a well-known result of the quantum theory of solitons (see, e.g. [15]) that the normal modes belonging to the continuum part of the spectrum of the differential operator in the equation for small disturbances describe the scattering states of the "mesons" off the "soliton". We can, therefore, use the normal modes to study scattering processes of photons and gravitons on the wormhole in zero approximation.

In the case $e^2+b^2=m^2$, we use a result [20] to obtain a rough estimate for the cross section σ_c for the capture of photons and gravitons by a wormhole:

$$(29) \qquad \sigma_c(\omega) \gtrsim 16\pi m^2 - 8\sqrt{2}\,\pi \frac{m}{\omega} ,$$

where ω is the frequency of the impact particle.

The result (29) is unusual in soliton theory. For a classical "lump" in flat spacetime, the effective potential in the differential equation for small disturbances around the soliton solution is always smooth and short range, because of the regularity of the soliton and the finiteness of its total energy. Thus, the high frequency mesons do not interact with the lumps.

In our case, the soliton is not a lump, but a hole. It has smooth, finite range potential barrier around it. However, not feeling the potential means, this time, falling into the hole.

If this effect survives in higher approximations, it can have interesting consequences: it will change the high energy behaviour of the theory.

Acknowledgements:

The author is indebted to J.-L. Gervais, G. Gibbons, W. Israel, K. Kuchař, M. Lüscher and H. Rumpf for valuable comments.

References

1. J.A. Wheeler: "Geometrodynamics", Adacemic Press, New York, 1962.

2. A. Salam, J. Strathdee: Phys.Letters 61B, 375 (1976).

3. G. W. Gibbons, M.J. Perry: Phys.Rev. D22, 313 (1980).

4. P. Hajicek: Nucl.Phys. B185, 254 (1981), Phys.Letters 106B, 77 (1981).

5. S. Coleman: Phys.Rev. D11, 2088 (1975).

6. G.W. Gibbons: Soliton States and Central Charges in Extended Supergravity Theories. Preprint, Cambridge, 1981.

7. L.D. Faddeev, V.E. Korepin: Phys.Reports 42, 1 (1978).

8. R. Rajaraman: Phys.Reports 21, 227 (1975).

9. R. Jackiw: Rev.Mod.Phys. 49, 681 (1977).

10. S. Coleman: in "New Phenomena in Subnuclear Physics". Ed. by A.Zichichi (Plenum, New York, 1977).

11. R. Arnowitt, S. Deser, C.W. Misner: in "Gravitation: an Introduction to Current Research". Ed. by L. Witten (Wiley, New York, 1962).

12. P. Hajicek: J. Math.Phys. 16, 518 (1975).

13. T. Regge, C. Teitelboim: Ann.Phys. 88, 286 (1974).

14. J.-L. Gervais, A. Jevicki, B. Sakita: Phys.Reports 23C, 281 (1976).

15. J.-L. Gervais, A. Jevicki, B. Sakita: Phys.Rev. D12, 1039 (1975).

16. D.G. Boulware: Phys.Rev. D11, 1404 (1975).

17. V. Moncrief: Phys.Rev. D9, 2707 (1974).

18. V. Moncrief: Phys.Rev. D12, 1586 (1975).

19. J. Bicak: Czech.J.Phys. B29, 945 (1979).

20. D.W. Olson, W.G. Unruh: Phys.Rev.Letters 33, 1116 (1974).

Phase Transitions in the Early Universe as a Consequence of a Generalized Theory of Gravitation

J.W. Moffat

Department of Physics
University of Toronto
Toronto, Ontario
Canada, M5S 1A7

Abstract: In the non-symmetric theory of gravitation the big-bang singularity is described as a critical point phenomenon (phase transition). The ordered (anisotropic) phase prior to the big-bang begins with a vacuum state at zero temperature. This scenario for the beginning of the universe explains the high degree of observed isotropy of the 3 K background radiation in the present universe and removes the "horizon problem" that occurs in the standard big-bang model.

I. Introduction

A generalized theory of gravitation based on a nonsymmetric fundamental tensor $g_{\mu\nu}$ has been extensively studied during the past three years[(1),(2)]. The physical motivation of the theory is to explain the stability of matter, using a basic U(1) gauge symmetry that occurs in the theory corresponding to the conservation of fermion number. Even if baryon and lepton number were observed to be violated experimentally, the conservation of fermion number F=B-L (B and L denote baryon and lepton number, respectively) may still be expected to hold true. This fundamental feature of nature has a natural dynamical explanation within the framework of NGT that does not contradict (local) gravitational experiments for the equivalence principle.[(2)]

In the following, I shall concentrate on one important consequence of NGT — the early universe cosmology. In the standard big-bang model based on general relativity (GR) there are four related problems:

1. How is matter created?
2. Why is the present universe isotropic?
3. How do we remove the horizon problem?
4. What is the nature of the big-bang singularity?

The problems generated by these four questions are problems of low-energy gravitational theory that may require a solution before an ambitious program of unification of the electroweak and strong forces with gravity can succeed. The solution may also shed light on the problems of quantum gravity. It appears highly unlikely, at this stage, that these questions can be answered using classical GR or some minor modification of GR.

In NGT the field structure, based on a nonsymmetric $g_{\mu\nu}$, leads to answers to the above four questions in a radically new way using the methods of critical phenomena physics. The basic results obtained by myself and D. Vincent will be published elsewhere[(3)]. Here, I shall discuss the analogies drawn between the results, obtained in ref.(3), and the physics of phase transitions that arise from diverse states of matter due to microscopic interactions involving many constituent particles,such as occur in liquids, magnets, superconductors and liquid crystals.[(4)]

2. Homogeneous, Anisotropic Solution of NGT

We shall adopt the perfect fluid description of matter with

$$T^{\mu\nu} = (\rho+p)u^{\mu}u^{\nu} - pg^{\mu\nu} , \qquad (2.1)$$

where ρ and p denote the mass density and pressure of the matter, and where the velocity four vector u^{μ} is for co-moving co-ordinates: $u^i=0$ $(i=1,2,3)$, $u^4=1$, $g_{\mu\nu}u^{\mu}u^{\nu} = 1$. Here $g_{\mu\nu}$ is nonsymmetric with $g_{\mu\nu}=g_{(\mu\nu)}+g_{[\mu\nu]}$ and $T=g^{\mu\nu}T_{\mu\nu}=\rho-3p$.

The four generalized Bianchi identities on the generalized Einstein tensor $G_{\mu\nu}$, give rise to the set of covariant conservation laws[2]

$$[(-g)^{\frac{1}{2}}g^{\alpha\nu}T_{\rho\nu} + (-g)^{\frac{1}{2}}g^{\nu\alpha}T_{\nu\rho}]_{,\alpha} + g^{\mu\nu}{}_{,\rho}(-g)^{\frac{1}{2}}T_{\mu\nu} = 0 . \qquad (2.2)$$

It has been shown[5] that if we split (2.2) into the two sets of equations

$$g_{(\rho\mu)}[(-g)^{\frac{1}{2}}T^{(\mu\nu)}]_{,\nu} + g_{(\sigma\rho)}[(-g)^{\frac{1}{2}}T^{(\mu\nu)}]\Gamma^{\sigma}_{(\mu\nu)} = 0 \qquad (2.3)$$

and

$$g_{[\rho\mu]}[(-g)^{\frac{1}{2}}T^{[\nu\mu]}]_{,\nu} + 2g_{[\mu\sigma]}(-g)^{\frac{1}{2}}T^{(\mu\nu)}\Gamma^{\sigma}_{[\nu\rho]}$$

$$+ g_{(\sigma\rho)}(-g)^{\frac{1}{2}}T^{[\mu\nu]}\Gamma^{\sigma}_{[\mu\nu]} + 2g_{(\mu\sigma)}(-g)^{\frac{1}{2}}T^{[\mu\nu]}\Gamma^{\sigma}_{[\nu\rho]} = 0 , \qquad (2.4)$$

then (2.3) yields the equations of motion of a test particle in NGT[1]:

$$\frac{d^2x^{\mu}}{ds^2} + \Gamma^{\mu}_{\alpha\beta}\frac{dx^{\alpha}}{ds}\frac{dx^{\beta}}{ds} = 0 . \qquad (2.5)$$

The field equations in NGT have been solved for a Bianchi Type I homogeneous, plane symmetric universe with the line element

$$ds^2 = dt^2 - \alpha(t)(dx^1)^2 - \beta(t)[(dx^2)^2 + (dx^3)^2]. \tag{2.6}$$

The solution is[6]

$$\begin{aligned} \alpha(t) &= R^2(t)(1 + \kappa^2 R(t)^{-4}) \\ \beta(t) &= R^2(t) \\ w(t) &= \kappa/R(t) , \end{aligned} \tag{2.7}$$

where $w(t) = g_{[14]}$ with all other components of $g_{[\mu\nu]} = 0$ and κ is a constant of integration. By substituting (2.1) into (2.3) and (2.4) using the relation $g^{\mu\nu}g_{\sigma\nu} = g^{\nu\mu}g_{\nu\sigma} = \delta^{\mu}_{\sigma}$, we get ($\dot{\rho} = \partial\rho/\partial t$)[3]:

$$\dot{\rho} = -(\rho + p)\frac{d}{dt}\,\ell n(-g)^{\frac{1}{2}} \tag{2.8}$$

and

$$\frac{\chi}{1-\chi}\,(\dot{\rho} - 2\frac{\dot{\beta}}{\beta}p) = 0 , \tag{2.9}$$

where $\chi = w^2/\alpha$. Eq.(2.8) has the canonical form of the energy conservation equation. Eq.(2.9) has two solutions: i) $\chi=0$ and ii) $p=-C_1R^4$ where C_1 is a constant. The first solution says that $w \equiv g_{[14]}=0$ which gives the Friedmann-Robertson-Walker (FRW) solution of GR. Substituting the second solution into (2.8) gives

$$\frac{d}{dt}(\rho R^3) = 3C_1R^6\dot{R} \tag{2.10}$$

and

$$\rho = \frac{3}{7} C_1 R^4 - \frac{C_2}{R^3} \quad , \tag{2.11}$$

where C_2 is a constant.

3. Phase Transitions and the Critical Point Singularity

Solving the field equations of NGT we get[3]

$$\dot{R} = \pm(8\pi G)^{\frac{1}{2}}\left(\frac{1}{7} C_1 R^6 - \frac{1}{3} C_2 \frac{1}{R}\right)^{\frac{1}{2}} \tag{3.1}$$

and

$$\ddot{R} = 4\pi G\left(\frac{6}{7} C_1 R^5 + \frac{1}{3} C_2 \frac{1}{R^3}\right) \quad . \tag{3.2}$$

Then $\dot{R}=0$ and $\ddot{R} > 0$ when $R=R^*=\left(\frac{7}{3}\frac{C_2}{C_1}\right)^{1/7}$ and this represents a <u>minimum</u> of R. A "bounce" occurs at $R=R^*$ and we choose $C_2/C_1 = \frac{3}{7}\kappa^{7/2}$ so that the bounce occurs at $R^* = \kappa^{\frac{1}{2}}$. We now find from (2.11) that

$$\rho(\kappa^{\frac{1}{2}}) = 0 \quad . \tag{3.3}$$

We have

$$T^{44} = \rho - p\frac{\chi}{1-\chi}$$
$$T^{ij} = -pg^{ij} \quad , \quad (i,j=1,2,3) \tag{3.4}$$

and for $R^* = \kappa^{\frac{1}{2}}$ this gives $T^{44} = -p = C_1\kappa^2$.

The anisotropy is defined by

$$A = \frac{|w|^2}{\kappa} = \frac{|\beta-\alpha|}{\kappa} \tag{3.5}$$

and for $R = \kappa^{\frac{1}{2}}$ we have A=1. The gravitational field energy $\sigma=\rho+p$ is <u>negative</u>. Solving for R from the field equations gives for $t \sim t_c$[3]:

$$R/R^* \sim \frac{1}{(t_c-t)^{\frac{1}{2}}} \tag{3.6}$$

so that

$$w \sim (t_c-t)^{\frac{1}{2}} \quad . \tag{3.7}$$

Also we find that

$$t_c = 0.675\left(\frac{3\kappa^{3/2}}{8\pi GC_2}\right)^{\frac{1}{2}} + t_o \quad , \tag{3.8}$$

where $t=t_o$ denotes the beginning of the universe.

We can now draw a striking analogy with the behaviour of a ferromagnetic system as solved by the mean field (or Landau) expansion of the free energy[4]. w plays the role of the <u>order</u> parameter and has the behaviour (3.7) for $t \leqslant t_c$. At the beginning of the universe $t=t_o=\kappa^{\frac{1}{2}}$, the temperature is zero and $w \neq 0$ with non-zero magnetization (anisotropy) and O(2) global spatial symmetry. Moreover at $t=t_o$ the universe is in a vacuum state with $\rho=0$. This vacuum state possesses stresses with $p < 0$ and is in "perfect order". Due to vacuum fluctuations the universe begins to expand and shortly after $t=t_o$ undergoes a <u>second order</u> phase transition at $t=t_c$

with a critical exponent $\beta = \frac{1}{2}$. For $t_o < t \leq t_c$, ρ increases as particles are created (due to the negative gravitational energy) and ρ goes through a singularity at $t=t_c$ in analogy with the static magnetic susceptibility of a ferromagnet. At $t=t_c$, ρ diverges symmetrically with the critical exponents $\gamma=\gamma'=2$. For $t > t_c$ we have $w=0$ corresponding to the disordered phase with O(3) symmetry (isotropy) described by the FRW solution of GR. In the vicinity of $t=t_c$ we cannot expect the universe to be in thermal equilibrium. A temperature duality with $T \to 1/T$ is expected to occur for $t \geq t_c$ as the universe cools down in the FRW model.

These results provide a logically consistent and self-contained solution to the issues raised by the discovery of the 3^oK microwave background and the peculiar nature of the initial conditions of the FRW universe in GR. The universe is initially singularity free (α, β and w are finite and non-vanishing for $t=t_o$) and a bounce mechanism is predicted at $t=t_o$, preventing the universe from collapsing below $t=t_o$. The vacuum state at $t=t_o$ with $\rho=0$ has stresses ($p < 0$) that cause an initial fluctuation, generating the expansion of the universe and the creation of matter. The "big-bang" singularity is the onset of the critical point phase transition at $t=t_c$ where the anisotropy A vanishes. Also at $t=t_c$ the horizon distance $d_H = \infty$, so that all of space is causally connected for a brief moment of time; this resolves the horizon problem.(3) The big-bang singularity becomes a "physical" type of singularity, since it corresponds to a phase transition (critical point) due to the correlations of a large number of fermions. Initially, in the ordered phase with $w \neq 0$, the net number of baryons minus anti-baryons $S^4=0$, as predicted by NGT for a flat universe(2)(3). Then for $t > t_c$ the antisymmetry of baryons and antibaryons can evolve in the FRW model as a signal of baryon number violation.

The long-range cooperativity at the beginning of the universe comes about through the interactions of particles with the $g_{[\mu\nu]}$ field which has the quantum numbers of a scalar field, $J = 0^+$ (7)-(9). Thus we see the emergence of the Goldstone theorem, since the symmetry breaking that occurs when we go from the higher O(3) symmetry to the anisotropic O(2) symmetry occurs for $w \equiv g_{[14]} \neq 0$ in the time interval $t_o < t < t_c$. Thus the $g_{[\mu\nu]}$ fields produce a collective mode at zero frequencies and long wavelengths generating long-range correlations.

We have used a mean field approximation (perfect fluid) to describe the early universe. It would be interesting to develop a microscopic theory of the long-range correlation in terms of a correlation function C with the order parameter

$$\langle w \rangle = \frac{1}{N_o} \langle \sum_i g^i_{[14]} \rangle \quad , \tag{3.9}$$

where N_o is the number of particles, <> represents an average over a statistical ensemble and $g^i_{[14]}$ describes the direction in space of a fermion particle.

Acknowledgement

This work was supported by the Natural Sciences and Engineering Research Council of Canada.

References

1 J.W. Moffat, Phys. Rev. D19, 3554 (1979).

2 For a review, see: J.W. Moffat, Proceedings of the VII International School of Gravitation and Cosmology, Erice, Sicily, edited by V. de Sabbata (World Scientific Publishing Co.). University of Toronto Report, December, 1981.

3 J.W. Moffat and D. Vincent, Can. J. Phys. (to be published).

4 c.f. H.E. Stanley, Introduction to Phase Transitions and Critical Phenomena, Oxford University Press, 1971.

5 R.B. Mann and J.W. Moffat, Can. J. Phys. 59, 1723 (1981).

6 G. Kunstatter, J.W. Moffat, and P. Savaria, Can. J. Phys. 58, 729 (1980).

7 R.B. Mann and J.W. Moffat, J. Phys. A14, 2367 (1981); Errata, to be published.

8 J.W. Moffat, Phys. Rev. D23, 2870 (1981).

9 R.B. Mann and J.W. Moffat, University of Toronto preprint, November 1981.

ON GROUP COVARIANCE AND SPIN MOTION IN GRAVITATIONAL THEORY

Leopold Halpern
Department of Physics
Florida State University
Tallahassee, Florida 32306

The general theory of relativity does not impose a particular local invariance group. The principle of equivalence is compatible with other groups than the Poincaré group.[2] A local invariance group is imposed by the theory of matter, particularly of spinors, but Lorentz and De Sitter covariant theories of non-gravitating spinor fields differ only by an immeasurably small constant of order of the reciprocal radius of the universe.[1]

The Poincaré group and the simple De Sitter and conformal groups differ in structure so strikingly that modifications of the physical laws beyond such small differences appear likely. The present paper outlines the fundamentals of a general relativistic theory with semi-simple invariance group. The deep-going influence of group structure and geometry on the physics is demonstrated. The approach differs fundamentally from presentations which appear only as an artifice to regain Poincaré group physics for the limiting value of a parameter. The only parameter used here in the example of the De Sitter groups, the radius of the universe, assumed to equal the reciprocal of the gravitational constant, is set equal to unity. The novel features of the theory (e.g. its cosmological fields) should either improve the description of nature (as it is hoped for the "spin mode of motion") or disprove the particular group covariance. This requires however careful analysis. The interrelation of natural phenomena seems to favour a semi-simple or simple group if group covariance really applied with adequate rigour; such a group is probably more sophisticated than the examples presented.

Most realization spaces of a group G are factor spaces with a subgroup. This is true for the De Sitter universe which are factor spaces of $SO(3,2)$ or $SO(4,1)$ with the subgroup $H = SO(3,1)$: $B = G/H$. Physical quantities are assumed to be related to either subgroups or to factor spaces of $SO(3,2)$ or subgroups. They manifest themselves in space time B by the natural projection π from the group manifold on B. For example the group manifold has a natural metric γ (Cartan-Killing metric). The projection π gives in the De Sitter case the metric g of the universe. In case of the conformal groups B is five dimensional and has to be projectively related to describe spacetime[4], similar as in the theory of Jordan and Thiry.[3]

The trajectories of particles are related to the projection of orbits of one dimensional subgroups of G or B. These are either geodesics on B or they perform for suitable initial conditions an additional mode of helical motion tentatively associated to an (unquantized) spinning particle motion.

The metric γ of the group manifold of dimension r fulfills source free Einstein equations with a cosmological member.

$$(1) \qquad R_{uv} - \frac{1}{2}\gamma_{uv} R + (\frac{r}{8} - \frac{1}{4})\gamma_{uv} = 0 \quad .$$

Therefore G, γ, H, π, B, g are the constitutents of a generalized Kaluza-Klein theory in which $P(G, \pi, G/H, H)$ form a principal fibre bundle.[4] The solution is thus globally defined without that B needs imbedding into G. The mathematical apparatus and its generalization to other solutions of Einstein's equations is outlined in section II.

II. Generalization and Discussion. Notation: A_S $(S = 1...r)$ is a basis of left invariant vector fields of G such that A_M $(M = K+1...r)$ represent the subgroup H. Capital or small indices a,b...k run from 1 to k, indices m...q from k+1 to r and indices s...z from 1 to r. This notation applies even to the summation convention. The A_S and their dual left invariant forms A^S fulfill:

$$(2) \qquad [A_U, A_V] = C^S_{UV} A_S \quad \text{and} \quad dA^S + \frac{1}{2} C^S_{UV} A^U \wedge A^V = 0 \; .$$

A connection exists on P with the A_E as horizontal and A_M as vertical vector bases. This connection is associated to a linear connection for the orthogonal subgroup H considered.

The right hand set of Eqs. (2) say that the torsion form (denoted by F^E) vainishes (first k eqs.) and the curvature form (denoted by $\mathring{F}^M$) equals: $\mathring{F}^M = -\frac{-1}{2} C^M_{EH} A^E \wedge A^H$ (last r-k eqs.). We shall denote these last (r-k) eqs. by $F^M = 0$ so that the right hand set of Eqs. (2) read: $F^S = 0$. The energy of $\mathring{F}^M$ is not positive definite; it interacts only with particles with nonhorizontal velocity components e_M (generalized charges which are conserved). Only such charges are admissible which draw no energy from $\mathring{F}^M$ they describe a helical motion on B.

$$(3) \qquad \ddot{x}^i + \{{}^{\,i}_{hk}\} \dot{x}^h \dot{x}^k = \mathring{F}^{Mi} \dot{x}^k e_M$$

"uncharged" particles move on geodesics. The charges can be eliminated from Eq. (3) to yield systems of differential equations of higher order. It is conjectured that in the De Sitter case their quantized version describes spin motion.

In the general case Eqs. (1) have source terms (localized masses) however such that the manifold retains the character of a principal fibre bundle with the same gauge group H and the same topology and a connection of the same type. The commutation relations of base vectors are now limited to:

$$(1a) \qquad [A_T, A_M] = C^S_{TM} A_S$$

but $[A_E, A_H]$ becomes arbitrary.
The F^S assume then more general values. The Lagrangian: is $\sqrt{g}\, R^{(r)}$ + cosmological member. The four dimensional form in the De Sitter case is:

$$(5) \qquad {}^{(4)} = \sqrt{g} \cdot (R^{(4)} + \frac{1}{4} \mathring{F}^{MiK} \mathring{F}_{MiK} + C)$$

varied with respect to g^{ik} and the potential of the gauge field (in local natural coords. A^M_h). It is bilinear in the F^S and can thus also be viewed as generalization of Einstein's teleparallelistic theory[5] in a form which is Vielbein-rotation covariant. The gauge field may mediate a spin-spin interaction. Spin and orbital angular momentum of many microscopic systems are hardly distinguishable. The puzzle is resolved by considering that the field of the binding forces in general interacts with A^M.

Applications of the formalism to theories including fermions and a theory in accord with Dirac's present large number hypothesis are outlined in Ref. (4) Emphasis in the theory is on nonlinear realizations and differential equations; microscopic Riemannian spaces may appear as obsolete as classical mechanics in the relativity age.

References

1. P.A.M. Dirac, Ann. Math., 36, 657 (1935).
2. L. Halpern, Journal of General Relativity, Vol. 8, N. 8 (1977), p. 623.
3. P. Jordan, "Schwerkraft and Weltall"; Y. Thiry, C.R. 226, p. 216 (1948).
4. L. Halpern, Proc. Meeting in honour of the 80th birthday of Dirac, New Orleans 1981. To appear in Int. J. Theor. Phys.
5. A. Einstein, Sitzungsberichte Der Preussischen, Akademie Physikalisch Mathematische Klasse.

QUANTUM GRAVITY ON A REGGE SIMPLICIAL NET AND ASYMPTOTIC SAFETY

M.Martellini
DAMTP, University of Cambridge, Cambridge (UK)
Istituto Nazionale di Fisica Nucleare, Sezione di Pavia, Italy *

Quantum gravity is a non-renormalizable theory. One could suppose that this is due to the fact that the Einstein's geometrodynamics is only an effective approximation. At very short distances it could be described by a so-called asymptotically safe theory, i.e. characterized only by a finite dimensional UV-critical surface. In order to show that, we shall write the pure quantum gravity on a Regge skeleton, regarding as dynamical variables the non-unitary vector fields $U^{(n)}$ which are normal to the 3-closed simplexes $T_3^{(n)}$.

In our formulation of the Regge calculus, the quantities which carry the curvature, the deficiency angles $\varepsilon^{(n)}$, are "essential coupling constants" which only in quantum theory, become functions of the lenghts of the 1-simplexes $T_1^{(n)}$. In fact, the request that the continuum Einstein's theory be independent of the type of simplicial net, leads us to interpret the $\varepsilon^{(n)}$ as "effective coupling constants" in the sense that $\varepsilon^{(n)} = \varepsilon^{(n)}[\ell_p^{(n)}]$, where $\ell_p^{(n)} = \lambda_p^{(n)} \ell$ and ℓ is the skeleton scale (it represents a "natural" UV-cutoff). At this level we recover the standard Regge's formulation, and the principal problem is to find how the $\varepsilon^{(n)}$'s vary with ℓ, when $\ell \to 0$ (the UV limit). Using the symbols of the Regge calculus, we have shown that the generating functional has an Ising-like gauge action with structure group $G = \overset{4}{\otimes} \mathbb{Z}_2$, namely

$$Z[J^{(n)}_{\alpha\beta}; K] = \int [dU^{(n)}_\alpha] \exp\Big\{ \frac{1}{2} \sum_{n\geq 1} \sum_{\alpha\beta\mu\nu=1}^{4} \Big[J^{(n)}_{\alpha\beta} \times \\ \times \delta_{\alpha\beta,\mu\nu} \big(U^{(n)}_\alpha U^{(n+1)}_\beta U^{(n)}_\mu U^{(n+1)}_\nu \big) + \delta_{\alpha\beta,\mu\nu} \times \\ \times K^{(n)}_{[\alpha\beta]} \big(U^{(n)}_\mu U^{(n+1)}_\nu \big) \Big] \Big\} \tag{1}$$

Here the saturations are performed w.r.t. the metric of the "interior" of the skeleton which is flat by construction, $\delta_{\alpha\beta,\mu\nu}$ is the alternating tensor, $U^{(n)}_\alpha$ are dimensionless vector-valued variables previously introduced and $J^{(n)}_{\alpha\beta}$

---

* Present address: Istituto di Scienze Fisiche, Università di Milano, Milano (Italy)

are strong coupling constants defined by

$$J^{(n)}_{\alpha\beta} = \frac{\varepsilon^{(n)}_{\alpha\beta}}{16\pi G_N \rho} \tag{2}$$

with $\varepsilon^{(n)}_{\alpha\beta}$ the deficiency angle which is common to the symplex $T_2^{(n)}$ (oriented by $U_\alpha^{(n)}$) and to the $T_2^{(n+1)}$ (oriented by $U_\beta^{(n+1)}$) and ρ is the "density" of the 2-closed simplexes $T_2^{(n)}$ (note that: $\ell \equiv 1/\sqrt{\rho}$). The renormalization theory can now be viewed as a description of how we can simultaneously change ℓ and the dimensionless coupling but nevertheless leave all physical results of the theory entirely unchanged. Namely, if the skeleton constant changes from ℓ to $\ell_p^{(n)} = \lambda_p^{(n)} \ell$ the strong coupling change from $J^{(n)}_{\alpha\beta}$ to $\hat{J}^{(n)}_{\alpha\beta}$, which may be written in the form

$$\hat{J}^{(n)}_{\alpha\beta}[\ell^{(n)}_p] = R^{\lambda_p}_{(n)}[J^{(n)}(\ell)] \quad ; \quad \lambda_p \equiv \lambda_p^{(n)} \tag{3}$$

Using the Migdal approximation in the hypothesis of performing successive decimations in the directions of the interior of the skeleton which we may approximate by a flat lattice, one gets that in $2+\nu$ -dimensions for an Ising like gauge action as in Eq. (1)

$$\hat{J}^{(n)}_{\alpha\beta} \sim \lambda_p^{\nu}\left\{ J^{(n)}_{\alpha\beta}(\ell) - \frac{1}{2}\ln\left(\lambda_p^{2+\nu-\beta} + \lambda_p^{1+\nu-\alpha}\right)\right\} , \quad \lambda_p \downarrow 0 \tag{4}$$

Eq. (4) shows that there exists a critical fixed point $J^{*}_{\alpha\beta}$ at large values of J, which is the same for each labelling n of the bones $T_2^{(n)}$. This fact ensures that all coupling constant $J^{(n)}_{\alpha\beta}$ approaches $J^{*}_{\alpha\beta}\left(= \frac{1}{2\nu}\ln[\lambda_p^{1+\nu-\beta} + \lambda_p^{\nu-\alpha}]\right)$ as $\lambda_p \downarrow 0$. A theory satisfying such a property has been called "asymptotical safety" by S.Weinberg. The surface formed by the trajectories (3), which is known as an UV-critical surface, is clearly one-dimensional. This means that it has no free-parameters aside from the one-dimensional constant M which merely defines our unit of mass that one may identify with the Planck mass. $M_p \sim 10^{19}$ GeV. This result has implicity assumed to neglect the atractor-UV-fixed point corresponding to the regular part of the free-energy, which in the continuum theory yields a cosmological constant. The dimensionality of the UV-critical surface would increase from one to two, if one wants to include also such a constant.